Designing with Xilinx® FPGAs

Sanjay Churiwala
Editor

Designing with Xilinx® FPGAs

Using Vivado

Editor
Sanjay Churiwala
Hyderabad, India

ISBN 978-3-319-82581-6 ISBN 978-3-319-42438-5 (eBook)
DOI 10.1007/978-3-319-42438-5

Softcover reprint of the hardcover 1st edition 2016

Printed on acid-free paper

This Springer imprint is published by Springer Nature
The registered company is Springer International Publishing AG Switzerland

Preface

The motivation for writing this book came as we saw that there are many books that are published related to using Xilinx software for FPGA designs. Most of these books are targeted to a specific version of Xilinx tools—be it ISE or Vivado or for a specific device. Xilinx makes two major releases of Vivado each year. Each release introduces significant new features and capabilities. Similarly, in each new device architecture, Xilinx makes significant enhancements. Hence, books written on any specific version of the software (or device architecture) get outdated very quickly. Besides, Xilinx anyways publishes its own set of documents which are updated with each major release of Vivado or FPGA architecture.

In this book, we have tried to concentrate on conceptual understanding of Vivado. These are expected to remain current through the current architecture of the tool chain. Our attempt has been that with a good conceptual understanding provided by this book, you will be able to understand the details provided in the user guides, which delve into the details of commands and options.

The Vivado software tool used for implementing a design on Xilinx's FPGAs has a lot of possible ways to read in a design. A user could describe the design in the form of HDL or "C" or make use of Xilinx-provided IP or use a third-party IP or the user could use his/her own HDL or "C" code as an IP to be used in multiple designs. A user could also describe the design using still higher level of abstractions using IP Integrator or SysGen. A design could also potentially use different types of inputs (for different portions of the design). You can use this book to understand the inherent strengths of the various modes of design entry. You can then decide which mechanism would be most suited for portions of the design. For the exact commands and syntax, you should refer to Xilinx documents. Our book provides a list of reference materials. Depending on which specific capability you plan to use, you can refer to the corresponding reference material.

Besides being useful to somebody who is new to Xilinx tools or FPGAs, the book may be found useful for those users who are migrating from ISE to Vivado. Vivado is conceptually very different from ISE. While ISE was mostly using proprietary formats for most of the flow, Vivado has moved on to industry standard formats. Users who have been long-time ISE users sometimes find it difficult to get used to Vivado. This book helps them get a good understanding of Vivado concepts, which should make it easier for them to transition to Vivado from ISE.

Though I've been involved in some of the user guides published by Xilinx, doing this book in my personal capacity allows me to deviate from the official stand also, wherever I wanted to, and share my real opinion. ☺

The most effective way to make use of this book is to not worry about reading the book from cover to cover. You can easily feel free to skip the chapters that deal with topics which your design does not have.

Hyderabad, India Sanjay Churiwala

Acknowledgments

I would like to express my gratitude to several of my colleagues and friends—within Xilinx and outside—who agreed to write the chapters on their areas of expertise and also reviewed each other's work. Each of these authors is highly knowledgeable in their respective areas. They took time out of their regular work to be able to contribute to this book.

I also thank my management chain at Xilinx, especially Arne Barras, Salil Raje, Victor Peng, and Vamsi Boppana—who were supportive of this work, even though this was being done in my personal capacity. I also thank the Xilinx legal/HR team, who provided me with the necessary guidance, permissions, and approvals to be able to complete this work, including usage of copyrighted material where relevant: Rajesh Choudhary, Lorraine Cannon Lalor, David Parandoosh, Fred Hsu, Cynthia Zamorski, and Silvia Gianelli. Amandeep Singh Talwar has been very helpful with figures and various aspects of the word processor. I often reached out to him, whenever I was having difficulty on either of these two aspects. Shant Chandrakar and Steve Trimberger helped me with specific items related to FPGA architecture. There are many more who have been supporting this actively.

I also thank my many teachers, colleagues, and seniors who have been teaching me so many things—that I could understand Semiconductor, EDA, and now specifically Xilinx FPGAs and Vivado. Over the last 23 years of professional experience in this field, there are just too many of such people that I dare not even try to name some, for the fear that I would end up filling up too many pages just with these names.

I also thank my family members. My immediate family members obviously adjusted with the fact that instead of spending time with them, I was working on this book. However, my entire extended family has been highly encouraging, by expressing their pride very openly at my past books.

And, I'm especially thankful to Charles Glaser of Springer, who is ever supportive of me working on any technical book. For this book, I also thank Murugesan Tamilselvan of Springer who is working through the actual processes involved in publication.

For me, writing continues to be a hobby that I cherish. And, once in a while, when I encounter somebody who identifies me with one of my books, the fun just gets multiplied many times for me. To anybody who has done this, I want to give a big "thanks" for encouraging me.

Contents

Chapter 1
State-of-the-Art Programmable Logic

Brad Taylor

1.1 Introduction

The FPGA or field-programmable gate array is a wonderful technology used by electronic system developers to design, debug, and implement unique hardware solutions without having to develop custom silicon devices. Xilinx is a semiconductor manufacturer of standard FPGA chips which are sold blank or unprogrammed to customers. The customers then program these devices to implement their unique systems. If a feature changes or a bug is discovered, the user can simply load a new program to the FPGA to create a new product or upgrade. This process can even continue after shipment in the form of firmware upgrades. The act of programming the FPGA is called configuration to distinguish it from loading any associated software programs. With modern FPGAs however, the line is blurring between hardware configuration and software programming.

All this programmability requires additional silicon area compared to hard ASIC (application-specific integrated circuit) implementations of the same logic. This is because in ASIC implementations the gates and wiring are fixed. This area cost penalty can be in the 1.5–10X range for FPGAs. However, the ASIC also must include the development cost and schedule which can be in the range of \$10–\$500 million dollars and can take several years with teams of hundreds of developers. With each generation of lithography, the cost to develop an ASIC increases. For these reasons, most medium-sized and smaller systems rely on a mix of FPGAs for customization along with standard ASIC or ASSPs and memories.

This revolutionary technology has impacted the electronic product development cycle for nearly all electronic devices since its introduction in the late 1980s.

B. Taylor (✉)
Santa Cruz, California, USA
e-mail: mail.brad.taylor@gmail.com

S. Churiwala (ed.), *Designing with Xilinx® FPGAs*,
DOI 10.1007/978-3-319-42438-5_1

1.2 The Evolution of Programmable Logic

The initial user programmable devices called *PLDs* (*programmable logic devices*) that were developed in 1978 by MMI could replace ten or so TTL gates and were one time programmable. This led to the reprogrammable PLDs based on EEPROM or EPROM technologies.

By 1985 advancing lithography enabled a new class of device, the FPGA. FPGAs introduced two important new architecture features: programmable routing to interconnect the increasing number of gates on a device and a programmable gate called a *LUT* or *lookup table* with an associated register. The initial devices from Xilinx contained up to a hundred *LUT* and flip-flop pairs in a basic logic element called a *CLB* or *configurable logic block*. Rather than using a permanently programmed EPROM or EEPROM memory, Xilinx FPGAs relied on CMOS memories to hold programming information. Figure 1.1 illustrates the technological improvement of modern FPGAs relative to the original Xilinx XC2064 which had 64 programmable logic cells.

The FPGA took its place as a central component in digital systems, replacing PLDs and TTL for implementing glue logic. In the 1990s new uses began to emerge for FPGAs, which were becoming more capable than just *gluing* I/O to processors. The emerging Internet became a growth driver for FPGAs with FPGAs being used for prototyping, initial deployment, and full-scale production of Internet switches and routers. By 2000 communications systems were the primary market for FPGAs. Other new markets for FPGAs also emerged for ASIC prototyping (Chap. 18) and high-performance DSP (digital signal processing) systems (Chap. 8). FPGAs also began to be used for implementing soft control processors such as the Xilinx MicroBlaze (Chap. 6) and PicoBlaze architectures.

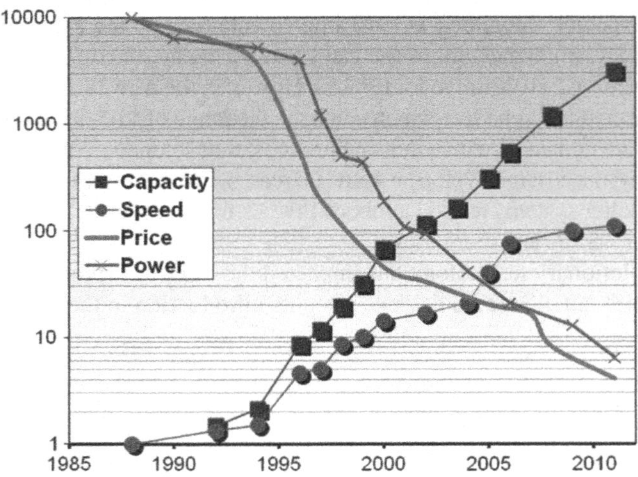

Fig. 1.1 FPGA evolution since the 1980s

The original FPGA architecture was a simple implementation of a programmable logic block. With each new generation, new programmable functions have been added along with hardening of some specific functions in order to reduce the cost or improve the performance of FPGAs in digital systems. These blocks continue to evolve in each generation. Many important functions have been added since the initial FPGAs including the following:

- Fast carry chains for high-speed adders and counters
- Small memories called distributed RAMs (or LUTRAMs)
- Block memories (BRAM or block RAMs)
- A hard RISC processor block based on the PowerPC
- Multi-Gigabit or MGT serial transceivers
- The DSP48 for digital signal processing
- Hard PCI blocks
- A complete system on chip (SoC) as a hard block in the FPGA in the Zynq family of FPGAs

The inclusion of hard blocks in FPGAs is driven by the trade-off between usage and cost. For customers which use these functions, value and performance are increased; however, if these hard blocks are not used, they are wasted space which can increase cost. Additionally these hard functions require significant software support to be useful to customers. For these reasons, hardening functions have been limited to those functions of clear value in important market verticals.

1.3 Current Applications for FPGAs

FPGAs find their usage in many applications today. Some of the most commonly used applications of FPGAs (and the reasons for FPGA being the sweet spot) include:

- ASIC prototyping: Chap. 18 covers more on this.
- Wired communications: For system development, while the standards themselves are evolving.
- Wireless communications: DSP in FPGAs is a major attraction for algorithmic computations.
- Video systems and machine vision: Implement software algorithms at higher speed and lower power.
- Industrial systems: Communication link between sensor nodes and robotic systems.
- Medical systems: I/O interfaces including A-to-D and D-to-A conversion.
- Automotive systems: Video processing (for driver assistance), field upgradability.
- Military and aerospace: Radio waveform processing and processing of huge amount of sensor data.
- Data center: Interfaces to SSD (solid-state disks), machine learning related algorithms.

1.4 Application Level System Architectures

The above applications in turn identify the need for the following system level usage, which might be applicable in multiple markets.

1.4.1 Glue Logic and Custom Interface IP

This was the original use case for early FPGAs. Typically the FPGA is used to interface a processor IC to a variety of I/O devices and memory-mapped devices. This use case requires low-cost FPGAs with plentiful I/O. Key features are combinatorial programmable logic nets, IOBs, and internal registers.

Often an application will require a custom interface such as an industrial interface or perhaps multiple interfaces such as USB. If these interfaces are not available in the user's SoC, they can be implemented in a companion FPGA.

1.4.2 Communications Switch

Multiple interfaces of various standards and performance levels such as 10G Ethernet are connected together via an FPGA implemented switch. These switches are common in Internet, industrial, and video networks.

1.4.3 I/O Stream Processing

FPGAs are ideal devices to connect to high-bandwidth real-time I/O streams such as video, radio, radar, and ultrasound systems. Often the system is used to reduce the high-native bandwidth of the I/O stream to levels manageable for a processor. For instance, a radio front end may sample A/D data at 1 GHz but after down conversion produces a more moderate rate of 10 MB/s. Conversely lower-bandwidth data may be up converted to a high-bandwidth I/O stream. Another example is a video system with a frame buffer which may be updated infrequently, but the video output stream is a real-time high-bandwidth stream.

1.4.4 Software Acceleration

An emerging FPGA system architecture allows software to be accelerated either with a companion FPGA attached to a high-end CPU or with an SoC-based FPGA such as the Zynq UltraScale+MPSoC (MPSoC). This acceleration will usually be accompanied by a significant power reduction per operation. In this use case, the FPGA is programmed on the fly to implement one or more cascaded software function calls on data in memory. The FPGA gates are compiled or derived from a common *C* language source which can be implemented either on the FPGA or on the CPU. This allows the FPGA to act as a high-performance library call for common software functions such as matrix inversion and deep neural networks.

1.5 FPGA Architecture

1.5.1 FPGA Architecture Overview

The primary function of the FPGA is to implement programmable logic which can be used by end customers to create new hardware devices. FPGAs are built around an array of programmable logic blocks embedded in a sea of programmable interconnect. This array is often referred to as the programmable logic fabric or just the *fabric*. At the edges are programmable I/O blocks designed to interface the *fabric* signals to the external world. It was this set of innovations that sparked the FPGA industry. Figure 1.2 shows a basic architecture of an FPGA.

Interestingly, nearly all the other special FPGA features such as carry chains, block RAM, or DSP blocks can also be implemented in programmable logic. This is in fact the approach the initial FPGAs took and users did implement these functions in LUTs. However, as the FPGA markets developed, it became clear that these special functions would be more cost effective as dedicated functions built from hard gates and later FPGA families such as the Xilinx 4 K series and Virtex began

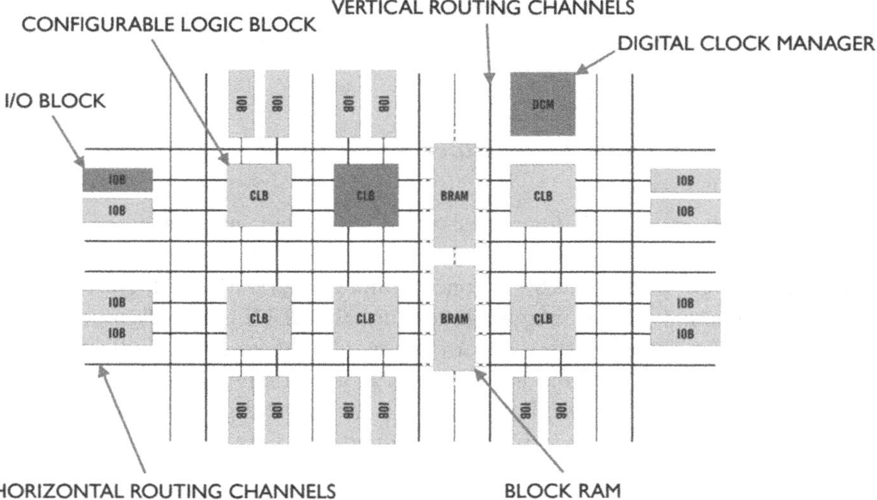

Fig. 1.2 Basic FPGA architecture

to harden these special functions. This hardening improved not only cost but also improved frequency substantially.

Within any one FPGA family, all devices will share a common fabric architecture, but each device will contain a different amount of programmable logic. This enables the user to match their logic requirements to the right-sized FPGA device. FPGAs are also available in two or more package sizes which allow the user to match the application I/O requirements to the device package. FPGA devices are also available in multiple speed grades and multiple temperature grades as well as multiple voltage levels. The highest speed devices are typically 25 % faster than the lower speed devices. By designing to the lowest speed devices, users can save on cost, but the higher performance of the faster devices may minimize system level cost.

Modern FPGAs commonly operate at 100–500 MHz. In general, most logic designs which are not targeted at FPGA architectures will run at the lower frequency range, and designs targeted at FPGAs will run in the mid-frequency range. The highest frequency designs are typically DSP designs constructed specifically to take advantage of FPGA DSP and BRAM blocks.

Sections below describe a high level overview of FPGA architectures. Please refer to Xilinx's data sheets and user guides for more detailed and current information.

1.5.2 Programmable Interconnect

Woven through the FPGA logic fabric is a set of wires which can be wired together to connect any two blocks in an FPGA. This enables arbitrary logic networks to be constructed by the user. The architecture of the interconnect wires varies from generation to generation and is hidden from the user by the tools.

1.5.3 Programmable Logic Block

An array of programmable logic blocks are embedded into the programmable interconnect. These are called CLBs (configurable logic blocks) in Xilinx devices. Today, each logic block consists of one or more programmable logic functions implemented as a 4–6-bit configurable lookup table (LUT), a configurable carry chain, and configurable registers. We use the word *configurable* to indicate a hard block which can be configured through the FPGA's configuration memory to be used as part of the user's logic. For instance, if the user design called for a register with a clock enable (CE), the register is configured to have the clock enable enabled and connected to the user's CE signal. Figure 1.3a through c illustrates the UltraScale CLB architecture, showing the CLB, LUT-flip-flop pair, and the carry chain structures.

The combination of a LUT, carry chain, and register is called a *logic cell* or *LC*. The capacity of FPGAs is commonly measured in logic cells. For instance, the largest Xilinx Virtex UltraScale FPGA supports up to 4 million LCs, while the smallest Spartan device contains as few as 2000 logic cells. Depending on usage, each logic cell can map between 5 and 25 ASIC gates. The lower number is commonly used for ASIC netlist emulation, while the higher number is achievable under expert mapping.

For Xilinx UltraScale devices, the CLB supports up to 8 × 6-input LUTs, 16 registers, and 8 carry chain blocks. Each 8-LUT can be configured as 2 × 5-LUTs if the 5-LUTs share common signals. For comparison purposes, Xilinx rates each 6-LUT as the equivalent of 1.6 LCs or Logic cells.

Embedded in the CLB is a high-performance look-ahead carry chain which enables the FPGA to implement very high-performance adders. Current FPGAs have carry chains which can implement a 64-bit adder at 500 MHz.

Associated with each LUT is an embedded register. The rich register resources of the FPGA programmable logic enable highly pipelined designs, which are a key to maintaining higher speeds. Each register can be configured to support a clock enable and reset with configurable polarity.

An important additional feature of the Xilinx CLB's 6-LUT is that it can configure to implement a small 64-bit deep by 1-bit wide memory called a distributed RAM. An alternate configuration allows the 6-LUT to implement a configurable depth shift register with a delay of 1–32 clocks.

1.5.4 Memory

Access to memory is extremely important in modern logic designs. Programmable logic designs commonly use a combination of memories embedded in the FPGA logic fabric and external DDR memories. Within the logic fabric, memory can be implemented as discrete registers, shift registers, distributed RAM, or block RAM. Xilinx UltraScale devices support two sizes of block RAM, 36-kbit RAMs and 288-kbit RAMs. In most cases the Xilinx tools will select the best memory type to map each memory in the user design. In some cases, netlists optimized for FPGAs will hand instantiate memory types to achieve higher density and performance.

Special forms of memory called dual-port memories and FIFOs are supported as special modes of the block RAMs or can be implemented using distributed RAM.

System memory access to external DDR memory (Chap. 5) is via a bus interface which is commonly an AXI protocol internal to the FPGA. UltraScale FPGAs support 72-bit wide DDR4 at up to 3200 MB/s.

In general, registers or flip-flops are used for status and control registers, pipelining, and shallow (1–2 deep) FIFOs. Shift registers are commonly used for signal delay elements and for pipeline balancing in DSP designs. Distributed RAMs are used for shallow memories up to 64 bits deep and can be as wide as necessary. Block RAMs are used for buffers and deeper memories. They can also be aggregated

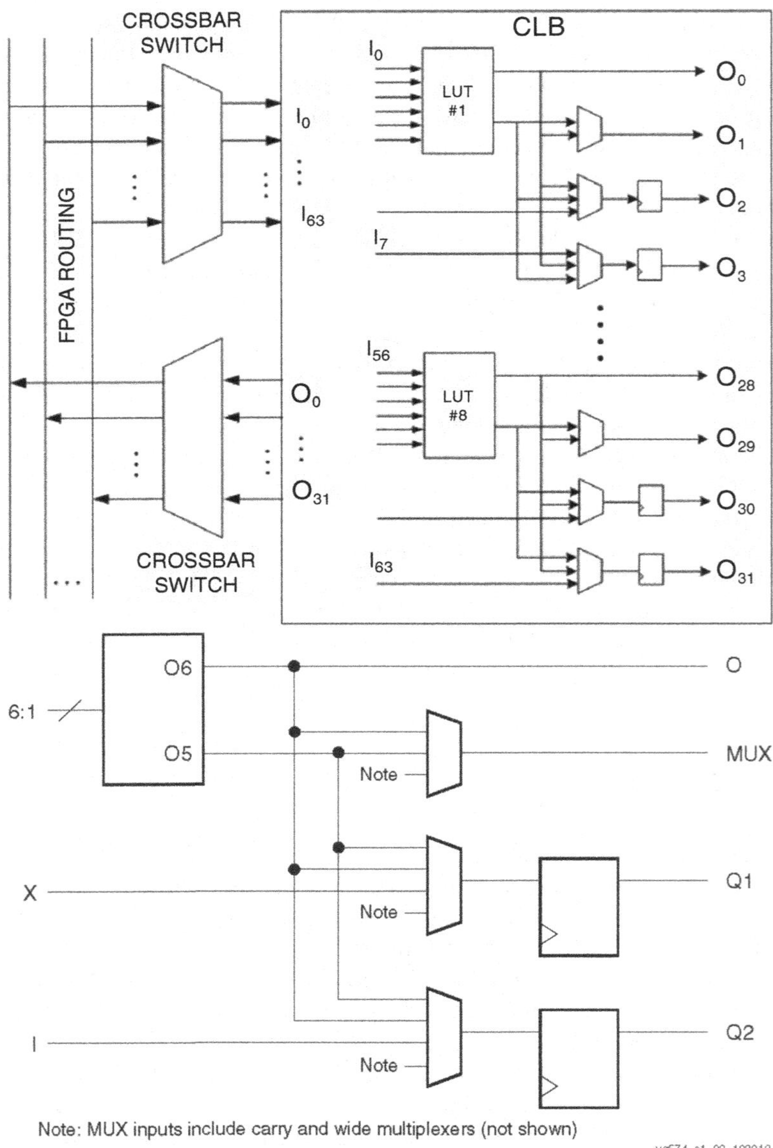

Fig. 1.3 (**a**) UltraScale CLB, (**b**) one of the eight LUT-flip-flop pairs from an UltraScale CLB, (**c**) carry chain paths

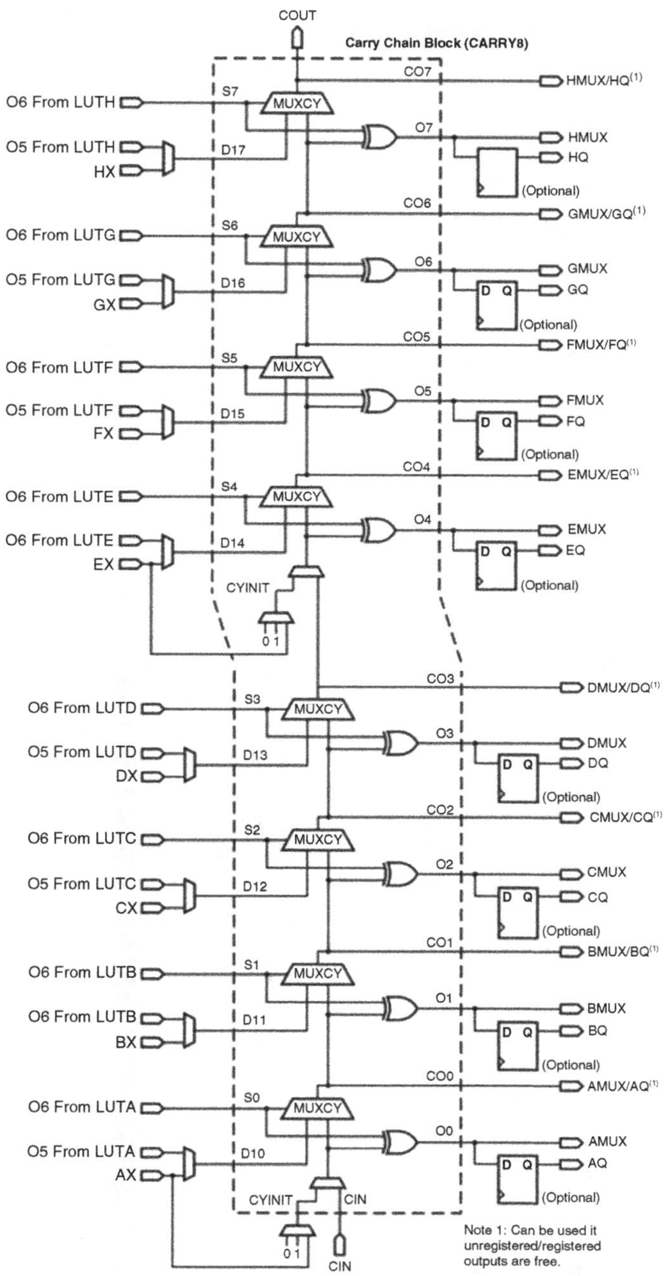

Fig. 1.3 (conrinued)

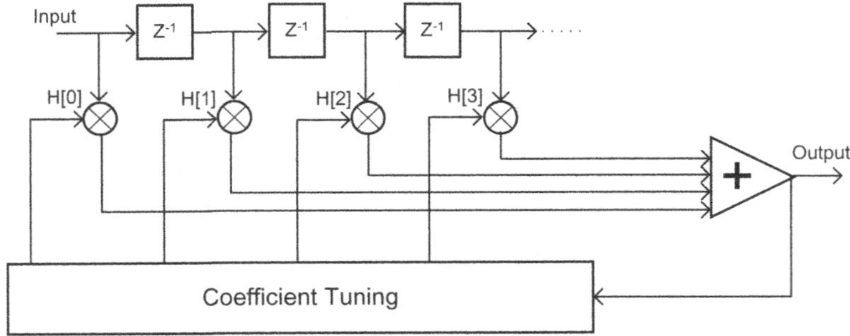

Fig. 1.4 DSP flowgraph

together to support arbitrary widths and depths. For instance, a 64-bit wide by 32 K-bit deep memory would require 64 block RAMs. Generally FPGAs contain around 1 36 K block RAMs for every 500–1000 logic cells.

1.5.5 DSP Blocks

Modern FPGAs contain discrete multipliers to enable efficient DSP processing. Commonly DSP applications build pipelines or flow graphs of DSP operations and data streams through this flow graph. A typical DSP filter called an FIR (finite impulse response) filter is shown in Fig. 1.4. It consists of sample delay blocks, multipliers, adders, and memories for coefficients. Interestingly this graph can be almost directly implemented as an FPGA circuit.

For filtering and many other DSP applications, multipliers and adders are used to implement the flow graph. Xilinx FPGAs contain a DSP block known as a DSP48 which supports an 18-bit × 25-bit multiplier, a 48-bit accumulator, and a 25-bit pre-adder. In addition up to four levels of pipelining can be supported for operation up to 500 MHz. The DSP48 supports integer math directly; however, 32-bit and 64-bit floating point operations are supported as library elements. A 32-bit floating point multiplier will require two DSP48s and several hundred LCs.

Xilinx tools will generally map multipliers and associated adders in RTL or HDL languages to DSP48 blocks. For highest performance however, designs optimized for DSP in FPGAs may use DSP48 aware libraries for optimal performance, power, and density.

1.5.6 Clock Management

Logic netlists almost universally require one or more system clocks to implement synchronous netlists for I/O and for internal operation. Synchronous operation uses a clock edge to register the results of upstream logic and hold it steady for use by downstream logic until the next clock edge. The use of synchronous operation allows for pipelined flow graphs which process multiple samples in parallel. External digital communications interfaces use I/O clocks to transfer data to and from the FPGA. Commonly, interface logic will run at the I/O clock rate (or a multiple of the I/O clock rate). Chapter 12 covers more on clocking resources available on Xilinx FPGAs.

1.5.7 I/O Blocks

One of the key capabilities of FPGAs is that they interface directly to external input and output (I/O) signals of all types and formats. To support these diverse requirements, modern FPGAs contain a special block called the I/O block or IOB. This block contains powerful buffers to drive external signals out of the FPGA and input receivers, along with registers for I/O signals and output enables (OE). IOBs typically support 1.2–3.3 V CMOS as well as LVDS and multiple industry I/O memory standards such as SSTL3. For a complete list, refer to the device datasheet. I/Os are abstracted from the user RTL and HDL design and are typically configured using a text file to specify each I/O's signaling standard.

UltraScale devices also include multiplexing and demultiplexing features in the I/O block. This feature supports dual data rate (DDR) operation and operation for 4:1 or 8:1 multiplexing and demultiplexing. This allows the device to operate at a lower clock rate than the I/O clock. For example, Gigabit Ethernet (SGMII) operates at 1.25 GHz over a single LVDS link, which is too fast for the FPGA fabric to support directly. The serial signal is expanded to 8/10 bits in the IOB interface to the fabric allowing the fabric to operate at 125 MHz.

I/Os are commonly a limited resource, and FPGAs are available in multiple package sizes to allow the user to use smaller lower-cost FPGAs with lower signal count applications and larger package sizes for higher signal count applications. This helps to minimize system cost and board space.

A primary application of FPGA I/Os is for interfacing to memory systems. UltraScale devices support high-bandwidth memory systems such as DDR4.

1.5.8 High-Speed Serial I/Os (HSSIO)

CMOS and LVDS signaling are limited in performance and can be costly in terms of power and signal count. For this reason, high-speed serial I/Os have been developed to enable low-cost, high-bandwidth interfaces. This evolution can be seen in the evolving PCI standard which has moved from low-speed 32-bit CMOS interfaces at 33 MHz to PCIe Gen3 with 1–8 lanes at 8 Gb/s lane. An eight-lane PCIe Gen3 interface can transfer 64 Gb/s of data in each direction. Xilinx UltraScale devices support up to 128 MGT (Multi-Gigabit Transceivers) at up to 32.75 Gb/s.

Within the FPGA, the HSSIO are interfaced directly to a custom logic block which multiplexes and demultiplexes the signals to wide interfaces at lower clock rates. This block also performs link calibration and formatting.

1.6 System on Chip

Current generation FPGAs now include an optional system on chip (SoC). These are available in the Zynq-7000 devices as well as the UltraScale + MPSoC devices. These SoCs include a state-of-the-art quad core ARM A53 application processor, an external DDR DRAM interface, internal memory and caching system, common I/O peripherals, and a set of high-bandwidth interfaces to the FPGA programmable logic.

The SoC is built using ASIC technology and is competitive with discrete embedded processors in cost and performance. It boots when powered up from an external flash memory. The processor is then available to load the FPGA design. While booting, the CPU boot code is optionally decrypted and authenticated enabling secure and safe embedded systems. Chapter 6 talks more about using these devices.

1.6.1 Operating System Support

The SoC system is capable of running bare-bones without an operating system or running a real-time operating system (RTOS) or embedded OSs such as Linux. It also supports asymmetric OSs where, for example, one core runs Linux and the other core runs an RTOS. This capability is ideal for embedded systems.

1.6.2 Real-Time OS Support

The MPSoC also includes a separate dual core ARM R5 processor. This processor is optimized for real-time applications and can optionally run in lockstep for high-reliability applications. The combination of the dual core R5 and the quad core A53 enables secure, high-reliability, real-time processing, while the A53 application processor executes application code. This combination is ideal for embedded, industrial, and military applications.

1.7 System Level Functions

In addition to the SoC and programmable logic array, FPGAs include system level functions for configuring and monitoring FPGAs.

1.7.1 System Monitor

For industrial and embedded applications, it is desirable to be able to monitor the voltage of system power supplies and various analog signals as well as the internal temperature of the FPGA. This allows the FPGA to detect if the power rails are within specified tolerance and allows the FPGA to know it is operating legally. For this reason and also for security reasons, FPGAs incorporate a small multichannel ADC (analog-to-digital converter). Chapter 16 covers more on system monitor.

1.7.2 Fabric Power Management

Before SoCs were introduced, FPGAs operated on a single power domain. Typically several voltages are required for the FPGA, the logic power supply, the auxiliary power supply, and the I/O power supplies. The FPGA fabric supports several features which allow the user to manage and minimize system power. FPGA fabric power consists of two types of power—static power which exists even if the device is not operating and dynamic power which is a function of clock rates and data activity. Static power is quite low at low temperatures but can rise to significant levels at maximum die temperatures. Additionally some speed and temperature grades have lower static power than others. The -2L speed grade is designed to operate at lower voltage levels enabling lower system power. The user has some flexibility to manage power by throttling fabric clocks if idle and by lowering die temperature using fan control.

1.7.3 SoC Device Power Management

The SoC devices introduce some additional flexibility in power management if the application allows for sometimes running in reduced functionality or idle modes. The Zynq-7000 devices support independent PS (processing system) and PL (programmable logic) power domains. Thus, if the PL is idle, its power supply can be removed. The MPSoCs support even finer-grained power domains and can be placed into low-power modes with only the R5s operating. This allows system power as low as 50 mW to be achieved for low-performance modes. Normal operation of the SoC would be in the 1–3 W range and the PL could be in the 2–20 W range.

1.7.4 Configuration

Both the PS SoC and the PL require configuration data to function. For the PS this is boot code, and for the PL, it is called the *bitstream* data. FPGAs will commonly include a dedicated block to configure the FPGA from various sources of bitstream data. Xilinx supports boot over JTAG, over a dedicated serial or parallel interface and from dedicated flash memory devices. In the SoC devices, configuration is supported by a configuration controller in the SoC. Optionally UltraScale devices can be booted over a PCIe interface, eliminating the cost of local flash storage and simplifying system level configuration data management.

1.7.5 Security

FPGA security is a relatively new concern, but modern devices contain multiple security features which are used to decrypt, authenticate, and monitor configuration data.

Encryption is used to obscure the configuration data which is stored in external memory devices. This is valuable to protect user IP (intellectual property) as well as to provide protection for passwords and keys embedded in the configuration data. FPGAs now store one-time programmable encryption key (of up to 256 bits) which is used to decrypt configuration data on the fly.

Today it is critical for system integrity to check configuration data for correctness before loading into the PL and SoC. The configuration controller optionally does this by first checking to see if the boot code or bitstream can be authenticated. The MPSoC devices support authentication of up to 4 K bits in hardware. If an authentication fails, the device cannot be booted. The bitstream is authenticated against a decryption key stored in external memory.

Additional features of MPSoC devices include tamper detection circuitry with clock, power, and temperature monitoring. This can be used to deter attacks based on operating the device outside of its legal operating conditions.

Within the Zynq UltraScale+PS, hardware is used to isolate various parts of the system. This can prevent the application code from overwriting the secure real-time code.

1.7.6 Safety

FPGAs are physical devices which are specified to operate under specific voltage and temperature conditions. They have a designed lifetime of 10 years of operation after which they may fail in various ways. During normal operation cosmic rays and alpha radiation from radioactive trace elements can *upset* device registers. For these reasons circuitry has been built into the FPGA to monitor configuration data changes due to upset or other effects. The FPGA configuration data is monitored for a digital signature. If this changes unexpectedly, a signal is raised which can reset the FPGA. Memories are particularly sensitive to upset, and all PL block RAMs and the large PS memories have added parity bits to detect a single event upset.

1.7.7 Debug

Getting a large FPGA to production is a challenging effort. In order to facilitate debugging a dedicated JTAG interface is provided on the FPGA and PS. This interface has access to the FPGA configuration system and the PS memory map. It can be used to download code and to test system level I/O interfaces. Cross-trigger circuitry is available to debug SoC software and PL hardware simultaneously. The PS also includes support for standard ICE debugging pods.

1.7.8 Performance Monitoring

The MPSoC includes a number of performance monitors which can check and measure traffic on the AXI interconnect. For the PL these performance monitoring blocks can be implemented in soft logic to monitor PL AXI events.

Chapter 2
Vivado Design Tools

Sudipto Chakraborty

The Vivado suite of design tools contain services that support all phases of FPGA designs—starting from design entry, simulation, synthesis, place and route, bitstream generation, debugging, and verification as well as the development of software targeted for these FPGAs.

You can interact with the Vivado environment in multiple ways. This includes a GUI-based interface for interactive users, as well as a command-line interface if you prefer to use batch mode. Vivado also supports a scripting interface with a rich set of Tcl commands. These multiple modes of interaction can also be combined in different ways to suit the exact needs of users. These are explained in detail below.

2.1 Project vs. Non-project Mode

There are two primary ways to invoke design flows in Vivado—using a project or a non-project mode. In the first case, you start by creating a project to manage all your design sources as well as output generated from executing design flows. When a project is created, Vivado creates a predetermined directory structure on disk, which contains folders for source files, your configurations, as well as output data. Once a project has been created, you can enter and leave the Vivado environment as needed, and each time you can start from where you left off, without having to start from scratch each time. The project-based environment supports the notion of *runs* which allow users to invoke design flows like synthesis and implementation. You are allowed to customize the design environment in multiple ways, and these configurations are also persisted in the project environment in the form of "metadata."

S. Chakraborty (✉)
Xilinx, Longmont, CO, USA
e-mail: sudipto@xilinx.com

S. Churiwala (ed.), *Designing with Xilinx® FPGAs*,
DOI 10.1007/978-3-319-42438-5_2

The directory structure created for a project is as follows:

```
<project>/
<project>.xpr                 : the main project file in text format
<project>.srcs/               : directory for sources local to a project
<project>.ip_user_files/      : directory for user accessible IP files
<project>.runs/               : directory for output data from synth/impl
<project>.sim/                : directory for output data from simulation
<project>.hw/                 : directory for hardware debug related data
<project>.cache/              : directory for locally cached data
<project>.ipdef/              : directory for local IP definitions
```

Not all of the above mentioned directories will always be created. For example, a Vivado project supports referring to design sources remotely from their original location or copying them locally inside the project directory structure, based on user preference. The <project>.srcs directory is only created if there are such local copies of source files present.

In the non-project mode, you interact more directly with the Vivado environment using lower level commands. This mode is called *non-project* because you do not directly create a project to get your design flows to complete. However, it is important to note that a project *object* does exist in this case also; it is created automatically to manage certain aspects of the design flows. This project object exists only in memory while your session is active and does not create the on-disk structure described above. Since there is no automatic persistence of data on disk, all data is maintained only in memory and available only during the current session. Hence, you need to make sure that all necessary output is generated before you exit the current non-project session of Vivado.

One interesting note here is that the project mode of Vivado is actually built on top of the non-project mode, as explained in Sect. 2.2.1.

2.2 GUI, Command Line, and Tcl

Vivado offers a fully interactive graphical user interface to allow you to more easily manage your design sources and go through all phases of the design flow. Vivado also supports doing all these operations in a non-GUI, command-line environment. The common connection between these two interfaces is the Tcl commands that drive Vivado. Almost all operations performed during the GUI mode end up issuing a Tcl command to the core Vivado *engine*. These commands are shown in the Tcl console in the GUI and are also captured in a journal file, which is typically located where Vivado was started from, and the file is named *vivado.jou*. When working in command-line mode, these Tcl commands can be issued directly without needing the presence of a GUI.

2.2.1 Interaction with Project/Non-Project

While it is common for GUI-based users to typically use the project mode, it is also possible to execute the flows in non-project mode while being in the GUI. Similarly, command-line users can choose to use either project mode or non-project mode.

The Tcl commands supported for project mode are higher level, macro style commands which perform many functionalities under a single command. The Tcl commands for the non-project mode, on the other hand, are more granular WYSIWYG (what you see is what you get) type of commands which only perform the specified operation, no more no less. Some project mode commands actually use many non-project commands internally to perform the desired operation. This explains the comment in Sect. 2.1 that project mode in Vivado is actually built on top of the non-project mode.

Scripts 1 and 2 are example scripts for project mode and non-project mode, which both perform the same operation, but the non-project script is more verbose since it uses more granular commands.

```
Script 1: Project mode example Tcl script
   create_project project_1
   add_files top.v child.v
   launch_runs -to_step write_bitstream impl_1
   close_project
```

```
Script 2: Non-Project Mode Tcl Script
   read_verilog top.v
   read_verilog child.v
   synth_design -top top
   opt_design
   place_design
   route_design
   report_timing_summary
   write_checkpoint top_routed.dcp
   write_bitstream top.bit
```

2.2.2 Runs Infrastructure

In the Script 1 and Script 2 examples, the *launch_runs* command is a macro command which is part of the Vivado *runs infrastructure*. This command internally creates a Tcl script which looks similar to the non-project example Script 2 and automatically launches this script with a new Vivado session to execute the flow.

Runs infrastructure allows managing the output products from design flow automatically. It also maintains status of the flow execution, such that if a design source file changes, it automatically lets you know that the previously generated output product is now *out-of-date* and if you relaunch the end step of a run, it automatically determines which previous steps need to be performed first and executes them automatically.

The *runs* infrastructure also allows parallel execution of independent portions of the design flows to complete the overall flow quicker. These parallel runs can be executed on multiple processors in the same host machine, or if a compute farm like LSF or GRID is available, the multiple runs can be executed on different host machines in the compute farm.

2.3 Overview of Vivado GUI

This section provides a high level overview of the Vivado GUI and some recommendation for first-time users. Vivado is designed based on a concept of layered complexity. This means using the tool for common tasks and designs is made as automated and easy as possible without having to have detailed knowledge of the tool. However, once you get more familiarized with the tool and want to use advanced features to control your design flows in a customized manner, Vivado allows you with higher control with finer granularity.

Vivado GUI and project-based mode is highly recommended for first-time users or those who want to get quickly up and running. Using the GUI makes it easy to use the various wizards (like *New Project* wizard) to get started. First-time users can leave all settings at default and let the tool decide best automatic options. There are several example projects included with Vivado which you can readily open and use to try out the design flows. If you want to try your own design, the only two minimum required pieces of input are an HDL file to describe the design and a constraint file to specify the timing intent and pin mapping of the in/out signals to specific FPGA pins.

Figure 2.1 shows the screenshot of the Vivado GUI with some of the key areas highlighted:

1. This area is called the *Flow Navigator*. It provides easy, single click access to the common design flow steps and configuration options.
2. This area shows the sources in the design. The first tab here shows a graphical view of the sources with modules and instance relationships. The other tabs in this area show other key aspects of design sources.
3. This area shows the properties of the items selected in the GUI.
4. This area shows the Tcl console in the GUI as well as various reports and design run related details.
5. This area shows the built-in text editor, information related to project summary, etc.
6. This is a view of a design open in the GUI, which is key to all the design implementation steps.

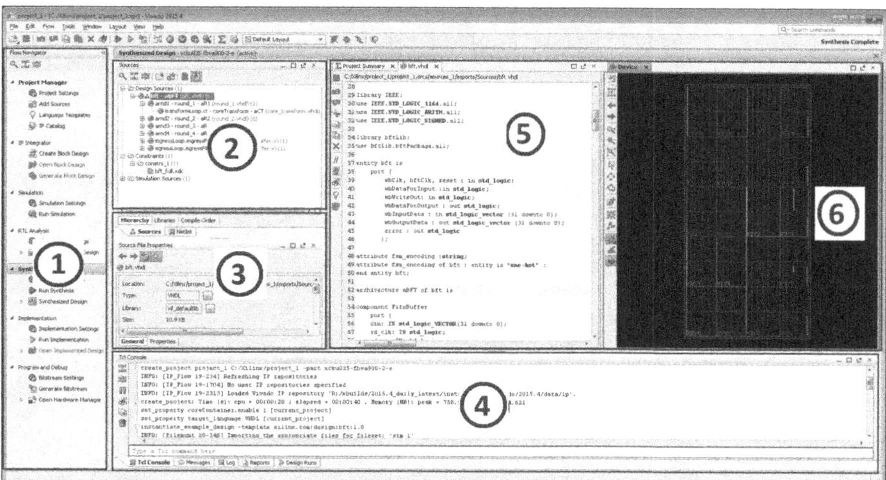

Fig. 2.1 Overall organization of Vivado GUI

Starting in the GUI and following the wizards make it easy to get started with the Vivado design flow. At the same time, as the various operations are being performed in the GUI, Vivado generates equivalent Tcl commands for those operations in the Tcl console area, as well as in the journal file as mentioned in Sect. 2.2. Using these Tcl commands, you can later customize the flow or build other similar flows.

Chapter 3
IP Flows

Cyrus Bazeghi

3.1 Overview

Intellectual property (IP) cores are fundamental criteria when selecting which FPGA vendor and specific part to choose for a design. IP provides an easy mechanism for incorporating complex logic in your designs, from high-speed gigahertz transceivers (GTs) to digital signal processors (DSPs) as well as soft microprocessors (MicroBlaze) to an embedded ARM system on a chip (SoC). Xilinx-provided IP have been optimized and tested to work with the FPGA resources including DPS, block RAM, and IO, greatly accelerating design development.

Most of the IP provided in the Vivado Design Suite have the license included, allowing the use of the IP in your designs. Some IP require a license to be purchased from Xilinx or an Alliance partner. IP licensing information is provided in the *IP Catalog* which will direct you to the appropriate web source.

The Vivado Design Suite includes the *IP Catalog* to deliver plug-and-play Xilinx IP as well as some third-party alliance partner IP. The catalog can be expanded with additional IP from third party IP developers or your own created IP. Your own IP could be created through:

- C/C++ algorithms compiled using the Vivado high-level synthesis (HLS) tool (see Chap. 10)
- Modules from system generator for DSP designs (MATLAB® from Simulink® algorithms) (see Chap. 8)
- Designs packaged as IP using the Vivado Design Suite IP Packager

C. Bazeghi (✉)
University of California Santa Cruz, San Jose, CA, USA
e-mail: cyrusbazeghi@outlook.com

S. Churiwala (ed.), *Designing with Xilinx® FPGAs*,
DOI 10.1007/978-3-319-42438-5_3

The Vivado IP Packager enables you to create plug-and-play IP which can be added to the extensible Vivado IP Catalog. The IP Packager is based on IP-XACT (IEEE Standard 1685), *Standard Structure for Packaging, Integrating, and Reusing IP within Tool Flows.*

After you have assembled a Vivado Design Suite project, the IP Packager lets you turn your design into a reusable IP module that you can then add to the Vivado *IP Catalog* and that others can use as a design source.

The IP Catalog is available either from within a Vivado Design Suite project or using a special *Managed IP* project. Both are available from the start screen.

The overall flow for IP consists of the following stages:

- Use the IP Catalog to find the IP required for the design.
- Customize the IP by specifying configuration options (produces an IP customization .xci).
- Generate the IP (Generate Output Products).
  - Copy files from the Vivado Design Suite installation area to the user-specified location or project.
  - By default include synthesizing the IP stand-alone (out-of-context).
- Instantiate the IP in designs (or in an IP integrator block design).
- Simulate.
  - Behavioral
  - Netlist
- Synthesize and implement.

3.2 IP Catalog

The *IP Catalog* (Fig. 3.1) provides a central and searchable place for all Xilinx-delivered IP, third party vendor IP, as well as user-created IP. To package RTL and constraints into a custom IP, the Vivado *IP Packager* is provided. The IP is grouped into categories either as a business segment such as *Communication & Networking*, *Automotive & Industrial*, by function such as *Digital Signal Processing* and *Debug & Verification* or by creator *Alliance Partner* and *User Repository*.

You can search by keyword, name, type, or function. You can expand to specify search options: case sensitivity (default is insensitive), toggle use of wildcards or use of regular expressions (default is neither), and match from start/match exactly/match anywhere (default).

Make sure to read the product guides for the IP cores that you plan to use in your design. The product guides provide important information on the functionality, requirements, use cases, and other details such as known issues for the IP which should be considered. The IP Catalog provides a convenient place to access the product guide, change log, product website, and any answer records for the IP.

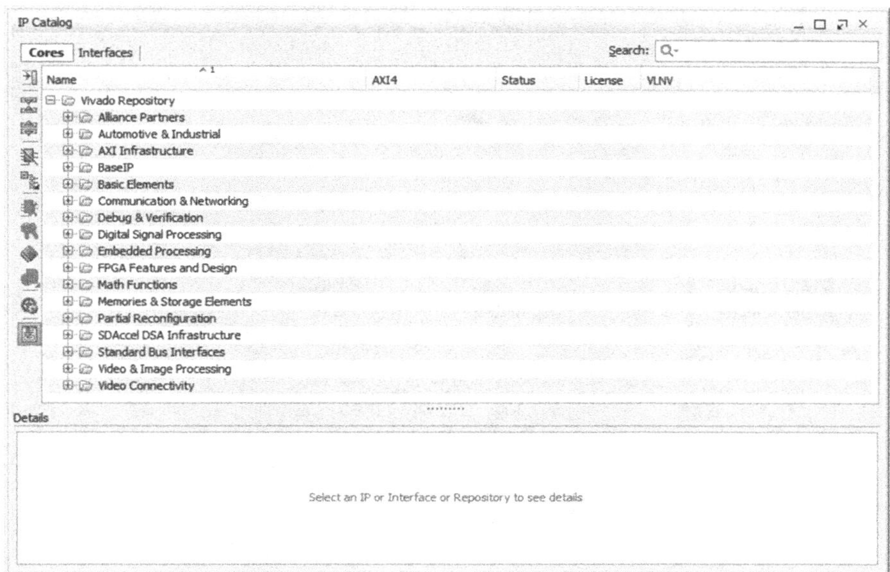

Fig. 3.1 Vivado IP Catalog

3.3 IP Customization

Each IP has many configuration options that can be set, for example, for a FIFO, width and depth, independent read and write clock, etc. (Fig. 3.2). A particular set of options for an IP is referred to as *customization* and will have a unique user-provided name. The customization options are encapsulated in the *IP_name.xci* file. Once an IP customization has been created, you can instantiate it in your design using the instantiation template (need to generate the output products to get this; see Sect. 3.4) as many times as required. Creating an IP customization does not add it to your design; you must instantiate it in your RTL for it to be used. You can create multiple customizations of the same IP, each with differing configuration options having a unique name.

There are three ways in which to create an IP customization:

- Managed IP Project (recommended)
- Directly from within a Vivado RTL project
- Using Tcl script

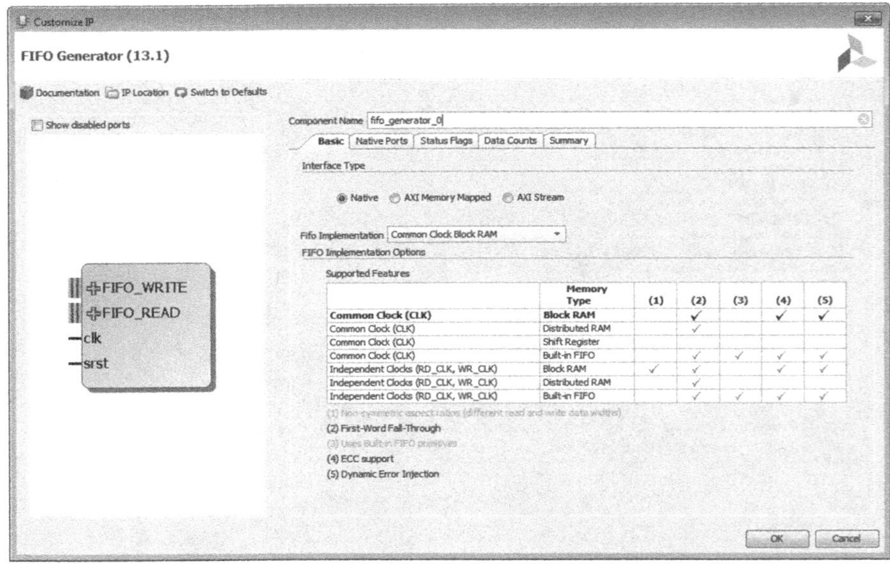

Fig. 3.2 Example of an IP customization GUI

Fig. 3.3 Icon for creating or opening a managed IP project

3.3.1 Managed IP Project

It is recommended when working with IP that you use a *Managed IP* project. This is a special Vivado project specifically for creating IP customizations. The same *IP Catalog* found in a Vivado RTL Project is provided to search for and customize IP. Each IP created is kept in its own folder with the user-provided name used during the customization. If you elect to use the *Core Container* feature (explained in Sect. 3.4.3), a single compressed binary file with the name given during customization with the extension of .xcix will be present. The IP folder or *Core Container* file are stored outside of the Managed IP Project directory. The IP folder or *Core Container* file can be copied into a revision control system along with other design sources. If electing to use a folder for the IP, it is recommended that you place the entire folder and all contents into revision control.

From the starting page of Vivado, select *Manage IP* (Fig. 3.3). You can either specify a new location on disk for a Managed IP Project or open an existing location.

3.3.2 *Within a Project*

You can elect to create IP customizations from directly within an RTL project. From the *Flow Navigator*, select *IP Catalog* and search/browse for the desired IP. During customization, by default the IP and associated files will be stored in the Vivado project directory structure. You can change this by the *IP Location* button, allowing you to specify a directory of your choice. This allows you to save the IP and its associated files outside of the Vivado project directory, similar as a *Managed IP* Project does. This is recommended when working with revision control systems.

3.3.3 *Tcl Script*

When creating an IP customization, Tcl commands are executed to create the IP and to apply the various options specified during customization. You can add these Tcl commands to your custom *Make* or script flow to create the IP on the fly. To compile your design, you would read in your RTL source, create the IP with Tcl commands, and proceed to synthesize or implement. The downside to this approach is that each time you build your project, the IP will have to be created and generated, which can be time consuming if there are many IP being used. Also, if the script was created with a previous version of Vivado, the IP might have changed the customization options which can result in errors being encountered.

3.4 IP Generation

Once you have customized your IP, it is recommended that you generate all the files that are needed by Vivado to make use of this IP. The generation of many IP can be done in parallel, including the synthesis of IP.

Generating output products or generating IP refers to these two stages (Fig. 3.4):

- Copying the IP files from the Vivado installation area
- Vivado processing the IP
  - Produce HDL based upon the customization options specified by the user.
  - Synthesize IP stand-alone (default).
  - Create simulation scripts for the IP for all supported simulators.

3.4.1 *Synthesis Options*

There are two options for synthesis of the IP at the end of customizing when presented with the *Generate Output Products* window (Fig. 3.4):

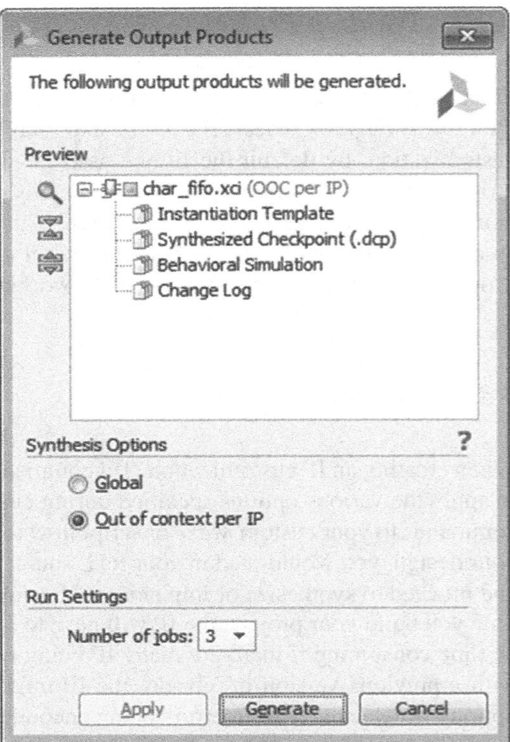

Fig. 3.4 Generate output products

- *Global*: The HDL files are created at this stage. However, the IP is synthesized along with the user HDL each time the design is synthesized.
- *Out-of-context per IP* (default): The IP is synthesized at this stage, just one time. During synthesis of the design the IP is seen as a black box. Later, during implementation, the netlist for all the IP will be combined with the top level netlist. This can greatly reduce the top level synthesis runtime during development.

There is no compelling reason to synthesize the IP globally. Since the Vivado tool places a *DONT_TOUCH* on the IP during synthesis, there are no cross boundary optimizations performed (explained in Sect. 3.6.2). Any cross boundary optimizations, such as propagation of constants, will be performed during implementation.

3.4.2 *Details of Generation*

Xilinx IP source files are delivered in the installation directory of the Vivado Design Suite. The IP consist of HDL (much of it encrypted) as well as scripts. The options specified during customization are processed by the scripts and HDL files are produced. Part of generation consists of copying all the static HDL files as well as the script-generated HDL files to the IP directory specified during IP customization.

Depending on the specific IP, different files will be copied from the install area to the user-specified IP folder (or to within the Core Container file) during generation. Possible types of files include:

- Static RTL files
- RTL files produced by scripts
- Instantiation templates
- Constraints
- Simulation files
- Testbench files
- Change log

Once this stage is completed, Vivado creates simulation scripts for all the supported simulators and places these scripts in the *ip_user_files folder* (*Managed IP* project) or *<project_name>.ip_user_files* (if IP created in a regular RTL project).

3.4.3 Core Container

The Core Container is a compressed binary file version of the IP folder *(.xcix)*. Vivado will read directly (not unzip to a temporary location) from the core container the files needed for synthesis, simulation, and implementation. Using this feature greatly reduces the number of files on disk needed for the IP. The Core Container is a standard ZIP file and can be opened with an appropriate utility, though modifying any of the contents is not supported and will likely cause issues with the use of the file.

To enable the Core Container for all new IP, go to the *Project Settings → IP→ General* tab and check the box *Use Core Containers for IP*. Alternatively, you can enable or disable the Core Container feature on a per IP basis. Select the IP in the *IP Sources* view, right click, and select *Enable Core Container* to enable. If enabled, you can right click and select *Disable Core Container*. This will switch between the IP being a folder on disk or an XCIX file and vice versa.

The *Core Container* is a complete representative of the IP you customized. All the files needed for Vivado are contained within. If using a project, the view of the IP will be identical regardless of using Core Container or not. You can open unencrypted HDL and constraint files, which will be listed in the IP Source view exactly the same as if not using Core Container. If outside of the Vivado project GUI, support files such as the instantiation template and simulation files can be extracted from the Core Container using the `export_ip_user_files` command. This will place them in the *ip_user_files* directory.

3.5 Using IP in Your Design

Using an IP is straightforward. If the IP was created in an RTL project, then simply use the provided instantiate template for either VHDL or Verilog to instantiate the IP in your design. The template is found in the *IP Sources* tab for the specific IP. If not using the Vivado GUI, the instantiation template can be found in the following locations:

- The IP directory
- In the *< project name>.ip_user_files* (IP created in an RTL project) or the *ip_user_files* directory (IP created in a *Managed IP* project)

If scripting your flow, read the IP using the `read_ip` command and pass the *< ip_name >.xci* or*<ip_name>xcix*. By referencing the *XCI/XCIX* file, Vivado will pull all required files in as needed, including HDL, *DCP* (if IP synthesized out-of-context), constraints, etc. If scripting a non-project flow, the IP must be fully generated.

Though you can use the IP DCP file in your flow, it is strongly recommended you use the *XCI/XCIX*. The reasons are:

- You can track the state of the IP going forward and can upgrade if you desire.
- During implementation, the IP XDC files will be processed in context of the entire netlist (see Sect. 3.6 for more details).
- If needed, you can make changes to the IP HDL (e.g., modify clocking resources).

3.6 IP Constraints

Most Xilinx IP come with constraint files (.xdc). They can contain physical constraints such as setting IO standards or locations and timing constraints, such as false paths. These two types can be mixed in the same file. The constraints are written as if the IP were the top of the design. The constraints are automatically scoped to IP instance(s). It is strongly recommended that you do not modify the constraints delivered by an IP.

There are two sources of constraints used by IP. If you're a user of IP from the available catalog, you need not worry about this distinction. However, this distinction would be of importance, if you are creating an IP of your own:

- XDC files created during generation of the IP and contained in the IP directory or the *Core Container*
- Constraints created by Vivado automatically during the processing of the IP

3.6.1 IP Delivered

There are three general types of XDC files which IP deliver:

- *OOC XDC*: This file provides the target frequency for any clocks which drive the IP for use during *out-of-context* synthesis. The file is not used during global synthesis or global implementation. The target frequency of each clock is set either during the IP customization via the GUI or by setting a property on the IP after it has been created. This special XDC file is always processed before all other XDC files that an IP may deliver. The file has an extension of `_ooc.xdc`. Only one such file is delivered per IP.

- XDC files marked with a *PROCESSING_ORDER* of *EARLY*: The constraints contained either provide clock objects or do not have any dependence, such as a clock provided from outside the IP. These files are typically named`<ip_name>.xdc`.
- XDC files marked with a *PROCESSING_ORDER* of *LATE*. The constraints contained have a dependency on an external clock(s) being provided. The clock(s) would come from the top level during global synthesis and during global implementation. During out-of-context synthesis and implementation, the `_ooc.xdc` provides the clock definition(s). These files have the extension of `_clocks.xdc`.

With the exception of the `_ooc.xdc`, not all IP deliver constraint files. It depends on the specific IP and its requirements. Typically larger and more complex IP deliver all three. Some IP may further break their constraints up, for example, putting the implementation specific constraints in one file and timing constraints in another file.

3.6.2 Vivado Delivered

During the processing of the IP, the Vivado tool creates additional constraints as follows:

- The `<ip_name>_in_context.xdc` file: This file is created for IP when using the default *out-of-context* synthesis flow, where IP is synthesized stand-alone. The *_in_context.xdc* is used during global synthesis, when the IP is a black box. After completing synthesis of the IP stand-alone, the IP is scanned to determine:

  - Does the IP output a clock object? Some IP produce clocks which could be used by other IP or by the user during global synthesis and implementation. The clocking wizard is an example of one such IP. The `_in_context.xdc` provides these clock definitions, which consist of `create_clock` command(s) which will put the clock object(s) on the boundary pin(s) of the IP, which will be a black box during global synthesis. This file is stored within the IP DCP file.
  - Does the IP contain clock or IO buffers? In this case a property is set on the respective boundary pin. With this property set on the IP black box, it will prevent global synthesis from unnecessarily inserting an additional one.

- The `dont_touch.xdc` file: Depending on the version of Vivado, this file might be seen being read in the global synthesis log (if the IP is synthesized globally) or in the IP *out-of-context* synthesis log (default). This file places a *DONT_TOUCH* on the boundary of the IP. This serves two purposes, to prevent the IP boundary pins from being optimized away and to prevent the IP hierarchy from being altered. This guarantees any constraints an IP delivers do not get invalidated. The *DONT_TOUCH* is removed at the end of synthesis. This allows constant propagation and cross boundary optimizations to be performed during implementation after the IP constraints have been applied. In later versions of Vivado, this may be done without the creation of the `dont_touch.xdc` file though messaging will be produced.

3.6.3 Processing Order

Constraints are processed in a specific order. For user XDC files, they are processed either in the order they are read using the `read_xdc` command in a script or in the order they are placed in the Vivado project (the compile order). IP are automatically processed along with the user files. Synthesis option decides which IP XDC files will be used: out-of-context (default) or global with the user HDL. The processing order is important, if your design has constraints that impact an IP or your design's constraints depend on the constraints of the IP. For such dependence, it is important that the dependent constraints are read later. Vivado provides you with an ability to process your XDC files before or after any IP delivered XDC by setting the *PROCESSING_ORDER*, though it is not common for users to change the *PROCESSING_ORDER* property for their XDC files. IP use this property to cause their various XDC files to come either before or after the user XDC files.

The order of XDC files during synthesis of the top level where the IP a black box, since it was synthesized out-of-context, is (default):

1. Any <ip_name>_in_context.xdc files
2. User files in the compile order set

The order of XDC files when the IP is set to use global synthesis is:

1. User files with the *PROCESSING_ORDER* property set to *EARLY*
2. IP files with the *PROCESSING_ORDER* property set to *EARLY*
3. User files with the *PROCESSING_ORDER* property set to *NORMAL* (default order for files is based upon the compile order)
4. IP files with the *PROCESSING_ORDER* set to *LATE*
5. User files with the *PROCESSING_ORDER* set to *LATE*

This is the same order that is used during implementation.

To see the order in which the XDC as well as the HDL files are processed, use the `report_compile_order` command. To see just the constraints, use the `-constraints` option. The output is organized into sections:

- HDL used during global synthesis
- HDL used during out-of-context IP synthesis
- Constraints used during global synthesis
- Constraints used during implementation

- Constraints used during IP out-of-context synthesis
- Constraints used during IP out-of-context implementation (results for this are for analysis only; to fully place and route logic for use, see the hierarchical design document)

3.7 IP Upgrade Decisions

Typically when moving to a new version of the Vivado Design Suite, the Xilinx IP in your design will most likely be out-of-date and it will be locked. Each release of Vivado only delivers one version of each Xilinx IP. Locked IP cannot be re-customized nor be generated. If you had fully generated your IP as recommended in Sect. 3.4, you can continue to use it *as is* since all the files needed for it are present.

You can review the change log and product guide for the IP in your design and determine if you wish to upgrade to the current version or not. The changes can vary from simple constraint changes, possible bug fixes, to the addition of new features. Some upgrades will require changes to your logic as the ports of the IP could change or the functionality might necessitate logic changes in your design.

The process of upgrading is straightforward. Select the IP either in the Vivado RTL project in the *IP Sources* area or in the *Managed IP* project, and right click and select *Upgrade IP*. Once upgraded, you can proceed to generation of the output products. For speed and convenience, you can upgrade multiple IP in parallel.

3.8 Simulation of IP

One of the biggest advantages of the Vivado Design Suite is the Xilinx IP are all delivered as HDL, enabling fast behavioral simulation. The HDL files needed for simulation are created during the generation of the output products. The files are all located in the IP folder or within the *Core Container* file. When using the *Core Container*, the simulation-related files are copied into the `ip_user_files` directory.

When using a Vivado RTL project and launching simulations from the GUI, all files required for simulating the IP are automatically sent to the simulator along with your HDL files. In addition to the integrated Vivado simulator (*XSIM*), Vivado can launch specific simulators from third parties. Chapter 11 covers more on simulation.

If you elect to simulate outside of the Vivado Design Suite, scripts are provided in the `ip_user_files` directory for each supported simulator. These scripts will reference IP files either from the IP directory or the `ip_user_files` as applicable depending on if you are using *Core Container* or not. The IP scripts can be incorporated into your own simulation scripts.

You can also use the `export_simulation` command to create a script to simulate your entire design, including the IP. The command can also copy all the simulation HDL files into the directory of your choice. This makes it very easy to start simulating your design.

Chapter 4
Gigabit Transceivers

Vamsi Krishna

4.1 Introduction to MGT (Multi-Gigabit Transceiver)

Xilinx® provides power-efficient transceivers in their FPGA architectures. Table 4.1 shows the maximum line rate supported by various transceivers for seven-series and UltraScale architectures. The transceivers are highly configurable and tightly integrated with the programmable logic resources of the FPGA. Because of very high degree of configurability of these transceivers, Vivado also comes with *GT Wizard*, which you can use to instantiate the transceivers with the right settings and connections. It is important to understand various characteristics of the transceivers. This will allow you to understand the system level implication of the configuration options that you chose in the Wizard.

4.1.1 Reference Clocks

The reference clock input is terminated internally with 50 Ω on each leg to 4/5 *MGTAVCC*. Primitives such as *IBUFDS_GTE2/IBUFDS_GTE3* are used to instantiate reference clock buffers. Advanced architectures like *GTHE3/GTHE4* support output mode of operation. The recovered clock (*RXRECCLKOUT*) from any of the four channels within the same *Quad* can be routed to the dedicated reference clock I/O pins. This output clock can then be used as the reference clock input at a different location. The mode of operation cannot be changed during run time.

The reference clock output mode is accessed through one of the two primitives: *OBUFDS_GTE3* and *OBUFDS_GTE3_ADV*. The choice of the primitive depends on your application.

V. Krishna (✉)
Xilinx, Hyderabad, Telangana, India
e-mail: vamsik@xilinx.com

S. Churiwala (ed.), *Designing with Xilinx® FPGAs*,
DOI 10.1007/978-3-319-42438-5_4

Table 4.1 Maximum line rate supported by various transceivers

	GTY	GTX	GTH	GTP
Seven series		12.5 Gb/s	13.1 Gb/s	6.6 Gb/s
UltraScale	32.75 Gb/s		16.375 Gb/s	

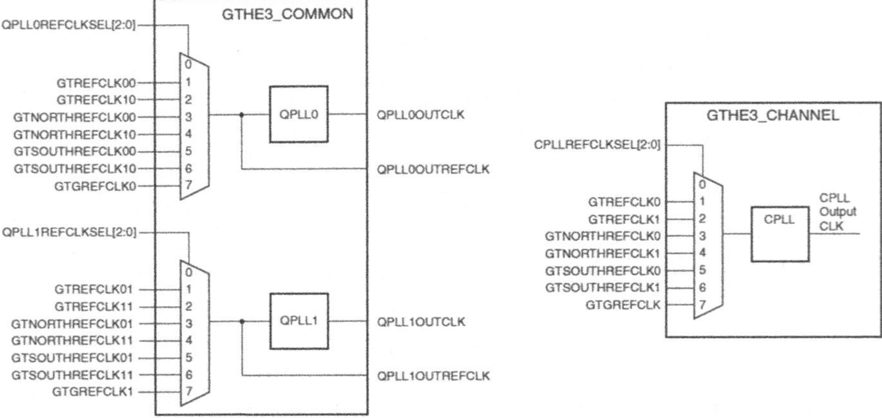

Fig. 4.1 QPLL and CPLL reference clock selection multiplexer

Figure 4.1 shows the detailed view of the reference clock multiplexer structure within a single *GTHE3_COMMON* primitive. The *QPLL0REFCLKSEL* and *QPLL1REFCLKSEL* ports are required when multiple reference clock sources are connected to this multiplexer. A single reference clock is most commonly used. In the case of a single reference clock, connect the reference clock to the *GTREFCLK00* and *GTREFCLK01* pins, and tie the *QPLL0REFCLKSEL* and *QPLL1REFCLKSEL* ports to *3'b001*.

Figure 4.1 also shows the reference clock multiplexer structure for the *GTHE3_CHANNEL* primitive. The *CPLLREFCLKSEL* port is required when multiple reference clock sources are connected to this multiplexer. For a single reference clock (which is the most common scenario), connect the reference clock to the *GTREFCLK0* port and tie the *CPLLREFCLKSEL* port to *3'b001*. Vivado will handle the complexity of the multiplexers and associated routing.

4.2 PLLs

4.2.1 CPLL

Each transceiver channel contains one ring-based channel PLL (*CPLL*). The internal channel clocking architecture is shown in Fig. 4.2. The *TX* and *RX* clock dividers can individually select the clock from the *QPLL* or *CPLL* to allow the *TX* and

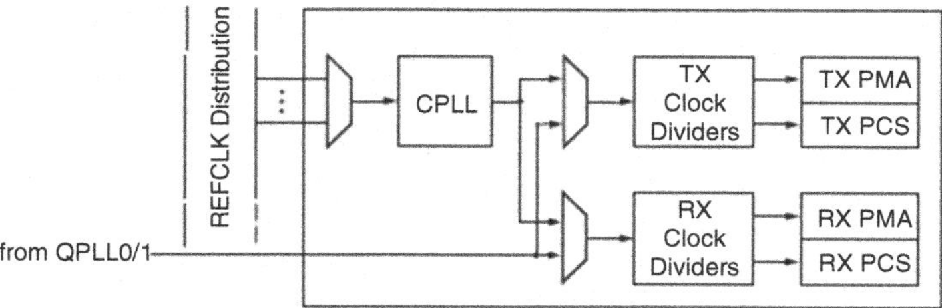

Fig. 4.2 Internal channel clocking architecture

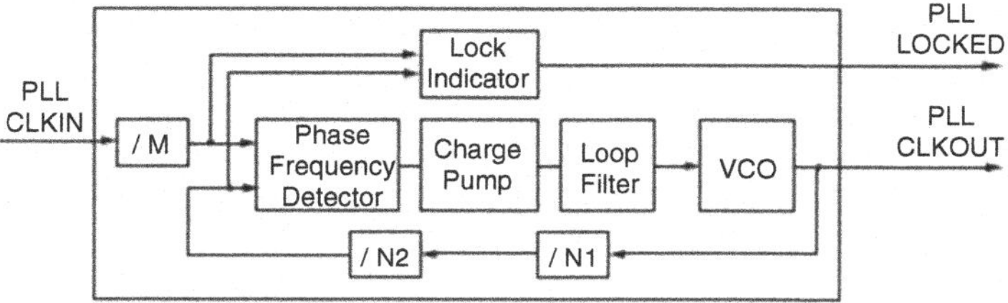

Fig. 4.3 CPLL block diagram

RX datapaths to operate at asynchronous frequencies using different reference clock inputs.

The *CPLL* outputs feed the *TX* and *RX* clock divider blocks, which control the generation of serial and parallel clocks used by the *PMA* and *PCS* blocks. The *CPLL* can be shared between the *TX* and *RX* datapaths if they operate at line rates that are integral multiples of the same *VCO* frequency. Figure 4.3 illustrates a conceptual view of the *CPLL* architecture. The input clock can be divided by a factor of *M* before feeding into the phase frequency detector. The feedback dividers, *N1* and *N2*, determine the *VCO* multiplication ratio and the *CPLL* output frequency. A lock indicator block compares the frequencies of the reference clock and the *VCO* feedback clock to determine if a frequency lock has been achieved.

4.2.2 QPLL

Each *Quad* contains one/two LC-based PLLs, referred to as the Quad PLLs (*QPLL0* and *QPLL1*). Either QPLL can be shared by the serial transceiver channels within the same *Quad* but cannot be shared by channels in other *Quads*. Use of *QPLL0/QPLL1* is required when operating the channels at line rates above the *CPLL* operating range. The *GTHE3_COMMON/GTHE2_COMMON* primitive encapsulates both the *GTH QPLLs* and must be instantiated when either *QPLL* is used. The *QPLL0/QPLL1* outputs feed the *TX* and *RX* clock divider blocks of each serial transceiver channel within the same *Quad*, which control the generation of serial and parallel clocks used by the *PMA* and *PCS* blocks.

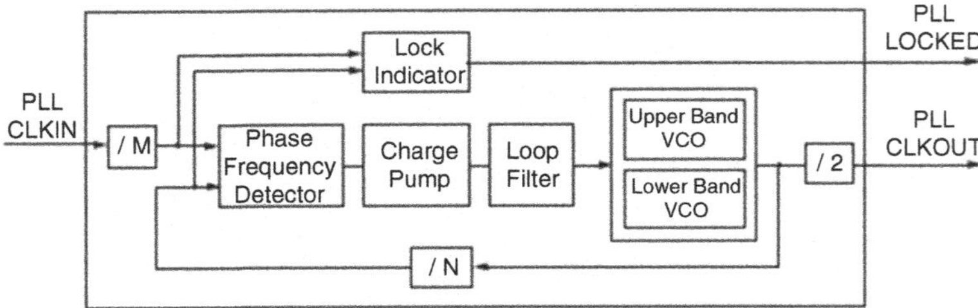

Fig. 4.4 QPLL block diagram

Figure 4.4 illustrates a conceptual view of the *QPLL0/QPLL1* architecture. The input clock can be divided by a factor of *M* before it is fed into the phase frequency detector. The feedback divider *N* determines the *VCO* multiplication ratio. The *QPLL0/QPLL1* output frequency is half of the *VCO* frequency. A lock indicator block compares the frequencies of the reference clock and the *VCO* feedback clock to determine if a frequency lock has been achieved.

4.3 Power Down

The transceiver supports a range of power-down modes. These modes support both generic power management capabilities as well as those defined in the standard protocols. The transceivers offer different levels of power control. Each channel in each direction can be powered down separately. Independent PLL power-down controls are also provided in transceiver.

4.4 Loopback

Loopback modes are specialized configurations of the transceiver datapath where the traffic stream is folded back to the source. Typically, a specific traffic pattern is transmitted and then compared to check for errors. Loopback test modes fall into two broad categories:

- Near-end loopback mode loop transmits data back in the transceiver closest to the traffic generator.
- Far-end loopback mode loop received data back in the transceiver at the far end of the link.

Loopback testing can be used either during development or in deployed equipment for fault isolation. The traffic patterns used can be either application traffic patterns or specialized pseudorandom bit sequences. Each transceiver has a built-in *PRBS* generator and checker.

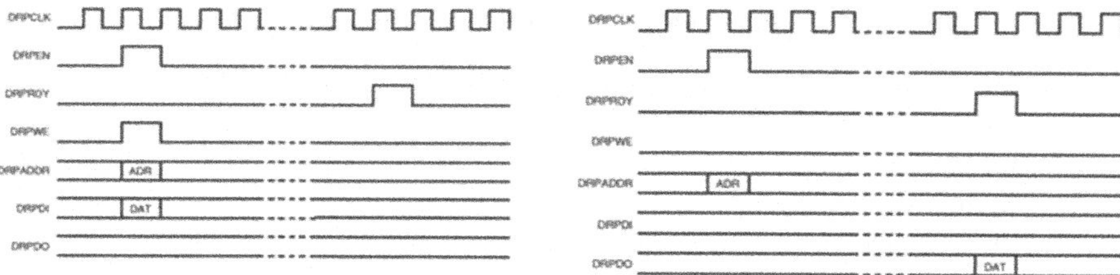

Fig. 4.5 DRP write operation (*left*) and DRP read operation (*right*)

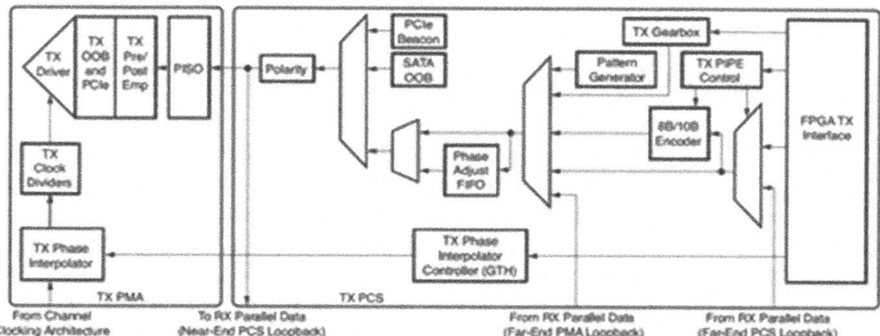

Fig. 4.6 TX block diagram

4.5 Dynamic Reconfiguration Port (DRP)

The dynamic reconfiguration port (*DRP*) allows the dynamic change of parameters of the transceivers and common primitives. The *DRP* interface is a processor-friendly synchronous interface with an address bus (*DRPADDR*) and separated data buses for reading (*DRPDO*) and writing (*DRPDI*) configuration data to the primitives. An enable signal (*DRPEN*), a read/write signal (*DRPWE*), and a ready/valid signal (*DRPRDY*) are the control signals that implement read and write operations, indicate operation completion, or indicate the availability of data. Figure 4.5 shows Write and Read timings. A new transaction can be initiated when *DRPRDY* is asserted.

4.6 Transmitter

Each transceiver includes an independent transmitter, which consists of a *PCS* and a *PMA*. Figure 4.6 shows the functional blocks of the transmitter. Parallel data flows from the FPGA logic into the FPGA *TX* interface, through the *PCS* and *PMA*, and then out of the *TX* driver as high-speed serial data.

Some of the key elements within the *GTX/GTH* transceiver *TX* are:

1. FPGA TX interface
2. TX 8B/10B encoder
3. TX gearbox
4. TX buffer
5. TX buffer bypass
6. TX pattern generator
7. TX polarity control
8. TX configurable driver

4.6.1 FPGA TX Interface

The FPGA TX interface is the FPGA's gateway to the TX datapath of the transceiver. Applications transmit data through the transceiver by writing data to the *TXDATA* port. The width of the port can be configured to be two, four, or eight bytes wide. The FPGA TX interface includes parallel clocks used in *PCS* logic. The parallel clock rate depends on the internal datawidth and the TX line rate.

4.6.2 TX 8B/10B Encoder

Many protocols use 8B/10B encoding on outgoing data. 8B/10B is an industry-standard encoding scheme that trades two bits overhead per byte for achieved DC balance and bounded disparity to allow reasonable clock recovery. The transceiver has a built-in 8B/10B TX path to encode TX data without consuming FPGA resources. Enabling the 8B/10B encoder increases latency through the TX path. The 8B/10B encoder can be disabled or bypassed to minimize latency, if not needed.

4.6.3 TX Gearbox

Some high-speed data rate protocols use 64B/66B encoding to reduce the overhead of 8B/10B encoding while retaining the benefits of an encoding scheme. The TX gearbox provides support for 64B/66B and 64B/67B header and payload combining. The TX gearbox has two operating modes. The external sequence counter operating mode must be implemented in user logic. The second mode uses an internal sequence counter. Due to additional functionality, latency through the gearbox block is expected to be longer.

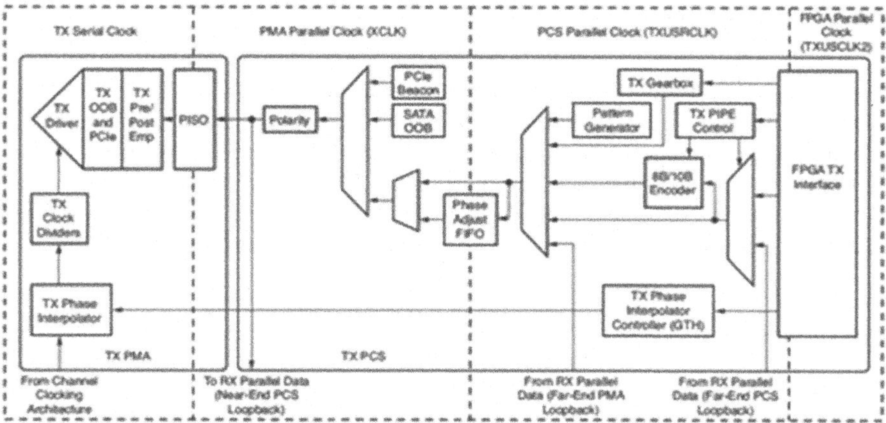

Fig. 4.7 TX clock domains

Table 4.2 TX buffer vs phase alignment

	TX buffer	TX phase alignment
Ease of use	The TX buffer is the recommended default to use when possible. It is robust and easier to operate	Phase alignment is an advanced feature that requires extra logic and additional constraints on clock sources
Latency	If low latency is critical, the TX buffer must be bypassed	Phase alignment uses fewer register in the TX datapath to achieve lower and deterministic latency
TX lane-lane Deskew		The TX phase alignment circuit can be used to reduce the lane skew between separate transceivers. All transceivers involved must use the same line rate

4.6.4 *TX Buffer*

The transceiver TX datapath has two internal parallel clock domains used in the *PCS*: the *PMA* parallel clock domain (*XCLK*) and the *TXUSRCLK* domain. To transmit data, the *XCLK* rate must match the *TXUSRCLK* rate, and all phase differences between the two domains must be resolved. Figure 4.7 shows the *XCLK* and *TXUSRCLK* domains.

The transmitter includes a TX buffer and a TX phase alignment circuit to resolve phase differences between the *XCLK* and *TXUSRCLK* domains. The TX phase alignment circuit is used when TX buffer is bypassed. All TX datapaths must use either the TX buffer or the TX phase alignment circuit. Table 4.2 shows the trade-off between buffering and phase alignment.

4.6.5 TX Buffer Bypass

Bypassing the TX buffer is an advanced feature of the transceiver. The TX phase alignment circuit is used to adjust the phase difference between the *PMA* parallel clock domain (*XCLK*) and the *TXUSRCLK* domain when the TX buffer is bypassed. It also performs the TX delay alignment by adjusting the *TXUSRCLK* to compensate for the temperature and voltage variations. The combined TX phase and delay alignments can be automatically performed by the transceiver or manually controlled by the user.

4.6.6 TX Pattern Generator

Pseudorandom bit sequences (*PRBS*) are commonly used to test the signal integrity of high-speed links. These sequences appear random but have specific properties that can be used to measure the quality of a link. The error insertion function is supported to verify link connection and also for jitter tolerance tests. When an inverted *PRBS* pattern is necessary, *TXPOLARITY* signal is used to control polarity.

4.6.7 TX Polarity Control

If *TXP* and *TXN* differential traces are accidentally swapped on the PCB, the differential data transmitted by the transceiver TX is reversed. One solution is to invert the parallel data before serialization and transmission to offset the reversed polarity on the differential pair. The TX polarity control can be accessed through the *TXPOLARITY* input from the fabric user interface.

4.6.8 TX Configurable Driver

The transceiver TX driver is a high-speed current-mode differential output buffer. To maximize signal integrity, it includes these features:

- Differential voltage control
- Precursor and post-cursor transmit preemphasis
- Calibrated termination resistors

4.7 Receiver

Each transceiver includes an independent receiver, made up of a *PCS* and a *PMA*. Figure 4.8 shows the blocks of the transceiver *RX*. High-speed serial data flows from traces on the board into the *PMA* of the transceiver *RX*, into the *PCS*, and finally into the FPGA logic.

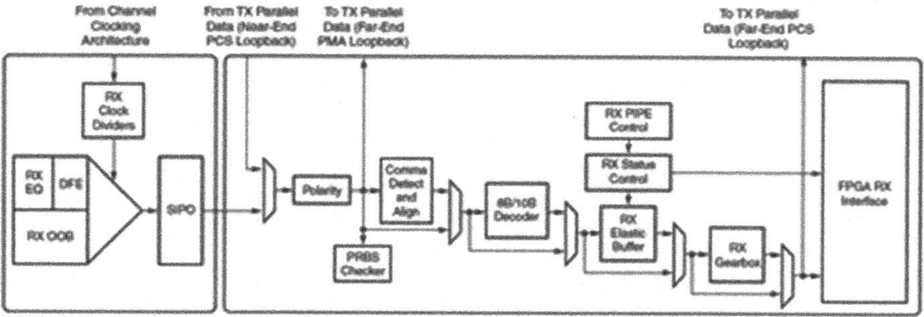

Fig. 4.8 RX transceiver block diagram

Some of the key elements within the transceiver RX are:

1. RX Analog front end
2. RX equalizer (DFE and LPM)
3. RX CDR
4. RX polarity control
5. RX pattern checker
6. RX Byte and Word Alignment
7. RX 8B/10B decoder
8. RX buffer bypass
9. RX elastic buffer
10. RX clock correction
11. RX channel bonding
12. RX gearbox
13. FPGA RX interface

4.7.1 RX Analog Front End

The RX analog front end (*AFE*) is a high-speed current-mode input differential buffer. It has these features:

- Configurable RX termination voltage
- Calibrated termination resistors

4.7.2 RX Equalizer (DFE and LPM)

A serial link bit error rate (*BER*) performance is a function of the transmitter, the transmission media, and the receiver. The transmission media or channel is bandwidth limited and the signal traveling through it is subjected to attenuation and distortion.

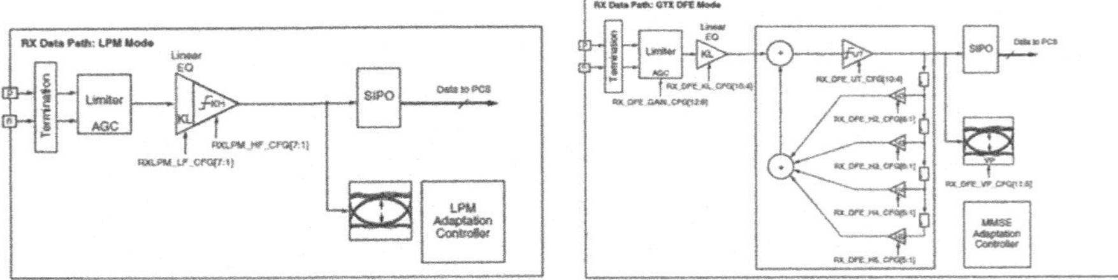

Fig. 4.9 LPM mode (*left*) and DFE mode (*right*) block diagram

There are two types of adaptive filtering available to the receiver depending on system level trade-offs between power and performance. Optimized for power with lower channel loss, the receiver has a power-efficient adaptive mode named the low-power mode (*LPM*), see Fig. 4.9.

For equalizing lossy channels, the *DFE* mode is available. See Fig. 4.9 for the transceiver. The *DFE* allows better compensation of transmission channel losses by providing a closer adjustment of filter parameters than when using a linear equalizer. However, a *DFE* cannot remove the precursor of a transmitted bit; it only compensates for the post-cursors. A linear equalizer allows precursor and post-cursor gain. The *DFE* mode is a discrete time-adaptive high-pass filter. The *TAP* values of the *DFE* are the coefficients of this filter that are set by the adaptive algorithm.

LPM mode is recommended for applications with line rates up to 11.2 Gb/s for short reach applications, with channel losses of 12 dB or less at the Nyquist frequency. *DFE* mode is recommended for medium to long-reach applications, with channel losses of 8 dB and above at the Nyquist frequency. A *DFE* has the advantage of equalizing a channel without amplifying noise and crosstalk. *DFE* can also correct reflections caused by channel discontinuities within the first five post-cursors in transceivers. *DFE* mode is the best choice when crosstalk is a concern or when reflections are identified in a single-bit response analysis.

Both *LPM* and *DFE* modes must be carefully considered in 8B/10B applications or where data scrambling is not employed. To properly adapt to data, the auto adaptation in both *LPM* and *DFE* modes requires incoming data to be random. Patterns with characteristics similar to *PRBS7* (or higher polynomial) are sufficiently random for auto adaptation to properly choose the correct equalization setting.

4.7.3 RX CDR

The RX clock data recovery (*CDR*) circuit in each transceiver extracts the recovered clock and data from an incoming data stream. The transceiver employs phase rotator *CDR* architecture. Incoming data first goes through receiver equalization stages. The equalized data is captured by an edge and a data sampler. The data captured by the data sampler is fed to the *CDR* state machine and the downstream transceiver blocks.

The *CDR* state machine uses the data from both the edge and data samplers to determine the phase of the incoming data stream and to control the phase interpolators (PIs). The phase for the edge sampler is locked to the transition region of the data stream, while the phase of the data sampler is positioned in the middle of the data eye.

The *PLLs* provides a base clock to the phase interpolator. The phase interpolator in turn produces fine, evenly spaced sampling phases to allow the *CDR* state machine to have fine phase control. The *CDR* state machine can track incoming data streams that can have a frequency offset from the local PLL reference clock.

4.7.4 RX Polarity Control

Similar to *Tx Polarity Control* (explained in Sect. 4.6.7), *RXPLOLARITY* (active *High*) input can be used to swap the *RXP* and *RXN* differential pins.

4.7.5 RX Pattern Checker

The receiver includes a built-in PRBS checker. This checker can be set to check for one of four industry-standard PRBS patterns. The checker is self-synchronizing and works on the incoming data before *comma* alignment or decoding. This function can be used to test the signal integrity of the channel.

4.7.6 RX Byte and Word Alignment

Serial data must be aligned to symbol boundaries before it can be used as parallel data. To make alignment possible, transmitters send a recognizable sequence, usually called a *comma*. The receiver searches for the *comma* in the incoming data. When it finds a *comma*, it moves the *comma* to a byte boundary so the received parallel words match the transmitted parallel words.

4.7.7 RX 8B/10B Decoder

If RX received data is 8B/10B encoded, it must be decoded. The transceiver has a built-in 8B/10B encoder in the TX and an 8B/10B decoder in the RX. The RX 8B/10B decoder has these features:

- Supports 2-byte, 4-byte, and 8-byte datapath operation
- Provides daisy-chained hookup of running disparity for proper disparity

- Generates K characters and status outputs
- Can be bypassed if incoming data is not 8 B/10 B encoded
- Pipes out 10-bit literal encoded values when encountering a not-in-table error

4.7.8 RX Buffer Bypass

Bypassing the RX elastic buffer is an advanced feature of the transceiver. The RX phase alignment circuit is used to adjust the phase difference between the *PMA* parallel clock domain (*XCLK*) and the *RXUSRCLK* domain when the RX elastic buffer is bypassed. It also performs the RX delay alignment by adjusting the *RXUSRCLK* to compensate for the temperature and voltage variations. Figure 4.10 shows the *XCLK* and *RXUSRCLK* domains, and Table 4.3 shows trade-offs between buffering and

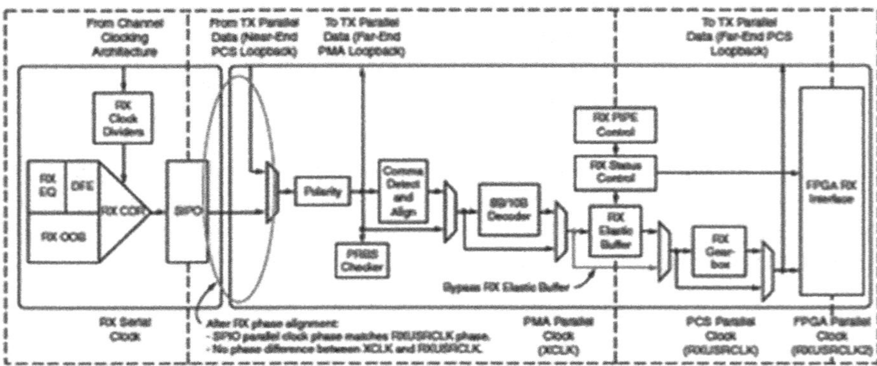

Fig. 4.10 RX phase alignment

Table 4.3 RX buffer vs phase alignment

	RX elastic buffer	RX phase alignment
Ease of use	The RX buffer is the recommended default to use when possible. It is robust and easier to operate	Phase alignment is an advanced feature that requires extra logic and additional constraints on clock sources
Clocking options	Can use RX recovered clock or local clock (with clock correction)	Must use the RX recovered clock
Initialization	Works immediately	Must wait for all clocks to stabilize before performing the RX phase and delay alignment procedure
Latency	Buffer latency depends on features use, such as clock correction and channel bonding	Lower deterministic latency
Clock correction and channel bonding	Required for clock correction and channel bonding	Not performed inside the transceiver. Required to be implemented in user logic

phase alignment. The RX elastic buffer can be bypassed to reduce latency when the RX recovered clock is used to source *RXUSRCLK* and *RXUSRCLK2*. When the RX elastic buffer is bypassed, latency through the RX datapath is low and deterministic, but clock correction and channel bonding are not available.

4.7.9 RX Elastic Buffer

The transceiver RX datapath has two internal parallel clock domains used in the *PCS*: the *PMA* parallel clock domain (*XCLK*) and the *RXUSRCLK* domain. To receive data, the *PMA* parallel rate must be sufficiently close to the *RXUSRCLK* rate, and all phase differences between the two domains must be resolved.

4.7.10 RX Clock Correction

The RX elastic buffer is designed to bridge between two different clock domains, *RXUSRCLK* and *XCLK*, which is the recovered clock from *CDR*. Even if *RXUSRCLK* and *XCLK* are running at the same clock frequency, there is always a small frequency difference. Because *XCLK* and *RXUSRCLK* are not exactly the same, the difference can be accumulated to cause the RX elastic buffer to eventually overflow or underflow unless it is corrected. To allow correction, each transceiver TX periodically transmits one or more special characters that the transceiver RX is allowed to remove or replicate in the RX elastic buffer as necessary. By removing characters when the RX elastic buffer is full and replicating characters when the RX elastic buffer is empty, the receiver can prevent overflow or underflow.

4.7.11 RX Channel Bonding

Protocols such as *XAUI* and *PCI Express* combine multiple serial transceiver connections to create a single higher throughput channel. Each serial transceiver connection is called one lane. Unless each of the serial connections is exactly the same length, skew between the lanes can cause data to be transmitted at the same time but arrive at different times. Channel bonding cancels out the skew between transceiver lanes by using the RX elastic buffer as a variable latency block. Channel bonding is also called channel *deskew* or *lane-to-lane deskew*. Transmitters used for a bonded channel all transmit a channel bonding character (or a sequence of characters) simultaneously. When the sequence is received, the receiver can determine the skew between lanes and adjust the latency of RX elastic buffers so that data is presented without skew at the RX fabric user interface.

4.7.12 RX Gear Box

The RX gearbox provides support for 64B/66B and 64B/67B header and payload separation. The gearbox uses output pins *RXDATA*[63:0] and *RXHEADER*[2:0] for the payload and header of the received data in normal mode. RX gearbox operates with the *PMA* using a single clock. Because of this, occasionally, the output data is invalid. The data out of the RX gearbox is not necessarily aligned. Alignment is done in the FPGA logic. The *RXGEARBOXSLIP* port can be used to slip the data from the gearbox cycle by cycle until correct alignment is reached. It takes a specific number of cycles before the bitslip operation is processed and the output data is stable. Descrambling of the data and block synchronization is done in the FPGA logic.

The RX gearbox operates the same in either external sequence counter mode or internal sequence counter mode.

4.7.13 FPGA RX Interface

The FPGA RX interface is the FPGA's gateway to the RX datapath of the transceiver. Applications transmit data through the transceiver by writing data to the *RXDATA* port. The width of the port can be configured to be two, four, or eight bytes wide. The rate of the parallel clock at the interface is determined by the RX line rate, the width of the *RXDATA* port, and whether or not 8B/10B decoding is enabled.

4.8 Integrated Bit Error Ratio Tester (IBERT)

The customizable LogiCORE™ IP Integrated Bit Error Ratio Tester (*IBERT*) core for FPGA transceivers is designed for evaluating and monitoring the transceivers. This core includes pattern generators and checkers that are implemented in FPGA logic and access to ports and the dynamic reconfiguration port attributes of the transceivers. Communication logic is also included to allow the design to be run time accessible through *JTAG*.

The IBERT core provides a broad-based Physical Medium Attachment (*PMA*) evaluation and demonstration platform for FPGA transceivers. Parameterizable to use different transceivers and clocking topologies, the IBERT core can also be customized to use different line rates, reference clock rates, and logic widths. Data pattern generators and checkers are included for each *GTX* transceiver desired, giving several different pseudorandom binary sequences (*PRBS*) and clock patterns to be sent over the channels. In addition, the configuration and tuning of the transceivers are accessible through logic that communicates to the dynamic reconfiguration port (*DRP*) of the GTX transceiver, in order to change attribute settings, as well as registers that control the values on the ports. At run time, the Vivado serial I/O analyzer communicates to the IBERT core through JTAG, using the Xilinx cables and proprietary logic that is part of the IBERT core.

Chapter 5
Memory Controllers

Karthikeyan Palanisamy

5.1 Introduction

External memory interface is an important component for majority of systems that are designed today. You have the option to choose various types of external memories depending upon the system requirements. The external memories required by you are supported through Xilinx Vivado IP catalog. Vivado provides options for you to configure various memory controllers as per your requirements. The performance of a memory subsystem would depend upon the access pattern to the memory, the electrical settings that are available, and the Vivado options. This chapter would go over the various types of memories that are available for you and the options that are available to configure the memory subsystem to get the required performance.

The on-chip memory available in an *FPGA* has increased over generations. The *FPGA* internal memories can be configured in various ways as per your requirements. The memory available in an *FPGA* can fully satisfy the memory requirements of a system or partially depending on the system requirements. Systems for which the memory requirements are more than that is available in an *FPGA* would opt for external memory. The type of memory used in a system will vary based on the system requirements. Various factors like *throughput*, storage requirements, power consumption, cost, and memory technology roadmap will go into selecting a memory interface.

Typically for an embedded system, a *DRAM* (dynamic random access memory) would be used as the external memory. *DRAM* memories are attractive due to the low cost per bit ratio, density, and availability. *DRAM* memories have evolved over time and the latest memories come with various power-saving features and also available at high data rates (3200 MT/s). A networking system would have a combination of

K. Palanisamy (✉)
Xilinx India, Hyderabad, India
e-mail: karthikeyan@yahoo.com

S. Churiwala (ed.), *Designing with Xilinx® FPGAs*,
DOI 10.1007/978-3-319-42438-5_5

DRAM memories and fast *SRAM* (static random access memory) or *RLDRAM* (reduced latency dynamic random access memory) memories. *RLDRAM* and *SRAM* memories are also feature rich and reach high data rates. *RLDRAM* and *SRAM* memories are expensive compared to *DRAM*, but they are attractive for applications that require low memory access times.

It is assumed that you have a good understanding of the memory technology, the roadmap, and the rationale behind choosing a memory for the system. The focus of the chapter will be on various options available for you and how you can set up the memory controller to achieve the system performance requirements.

Every memory IP has a product guide associated with it. The product guide will describe the IP in detail. You are encouraged to read the product guide for a particular IP to get in-depth knowledge of the IP. For every *FPGA* generation, there is also a PCB design user's guide which has many details regarding the PCB considerations that has to be taken into account for a memory interface.

5.2 Getting Started

Xilinx memory solutions are part of Vivado IP catalog. You can generate various memory controller designs by selecting the IP cores available in the IP catalog. The following memory IPs are generally available for every generation of *FPGAs*:

- *SDRAM*: *DDR3* and *DDR4*
- *SRAM*: *QDRII+* and *QDRIV*
- *RLDRAM*: *RLDRAM-3*
- *LPDDR*: *LPDDR3* and *LPDDR4*

The variants for memory devices will vary for every generation of *FPGAs* based on the memory roadmap and availability. In this chapter only the memory that is prevalent will be discussed in detail. For example, in *SDRAM* the chapter will go into details for *DDR4*. Most of the *DDR4* concepts are applicable for *DDR3, LPDDR3*, and *LPDDR4*.

5.2.1 Design Generation

Through the IP catalog in Vivado, you will be able to invoke the memory wizard tool for a given IP. The wizard will have multiple options to configure the IP. The options are split into "basic options" and "advanced options." The basic options would be used to configure the following:

- *Controller option*: The memory controller is split into two parts: the physical layer and the controller. The controller converts the user commands into the particular memory protocol commands. The controller will also have features built in to improve the efficiency. The physical layer is responsible for initializing the memory and performing calibration. The main function of calibration is to capture read data reliably and to send write data with enough margin to the memory. Calibration is required to make sure the memory interface works reliably over a range of process voltage and temperature. You can choose to generate a design with memory controller and physical layer or just the physical layer. Usually a physical layer-only design is generated for use in cases in which a system requires a custom controller to be used with the physical layer. The physical layer-only option is usually only applicable to *SDRAM* designs which has complex controller functions.

- *Clocking option*: The memory controller frequency of operation will be chosen here. There will be an option to select the clock input to the memory controller.
- *Memory device*: You can choose the memory device as per your requirements. The menu will have multiple memory devices and configurations for you to choose from. If a particular memory type is not available, the tool provides options to generate a custom memory part for you. Other options like the data width, read and write latencies (if applicable), and burst length would be chosen here. For certain memory types, some pins like chip select or data mask are optional and that selection would be done here. Finally if applicable for certain memory types, the option to have ECC will also be provided here.

With the basic options described above, you can generate a memory controller that will satisfy your needs. You can use the advanced option to further customize the memory controller. The advanced options would vary by the memory controller type. In general there would be options to select the following:

- By default the controller is configured for efficiency based on the default options. You can select switches that would improve efficiency for your traffic pattern.
- Provides advanced option for you to choose the input clock configuration.
- Option to provide debug hooks and bring the status signals to the Vivado Labtools Hardware Manager for easy debug. All the memory controllers will come up with a status viewer from Vivado Labtools by default. The status viewer will display the calibration status, read/write margin, and other relevant information.
- Advanced options provided to speed up simulation with behavioral models.
- Option to generate additional clocks that is synchronous to the memory controller clock. The additional clocks would be useful for a system that needs to clock other blocks that are synchronous to the chosen memory controller.
- Options to enable other controller-specific advanced features. For example, self-refresh feature in *DDR4* designs.

5.2.2 Pin Planning

The pin planning for the memory controller would be done in the main Vivado I/O Pin planner. To access the Vivado I/O planning environment, you would have to open the elaborated RTL design or the synthesized design. Once the design is opened, the I/O planning layout option can be chosen from the menu and pin selection can be done. A default pin out would be preloaded for the memory controller in the I/O planner. You can go with the default or choose your own custom pin out. In custom pin out, you have the option to do byte-level or pin-level selections. You can also read in an existing pin selection through the XDC file and use it for the IP pin out in I/O planner.

5.2.3 Example Design

The memory controller solutions have an example design associated with it. The example design can be invoked by right clicking on the generated IP in the Vivado console and choosing the option "Open IP Example Design."

The example design (apart from the IP files that was generated) will have the necessary files for simulation and implementation. The example design is useful for you to get a quick start with implementation and simulation of the generated memory controller. The example design can also be used as an instantiation template when you integrate the memory controller IP in your system. A traffic generator that can send in different traffic patterns depending upon your options will be part of the example design. The traffic generator can generate patterns like PRBS23 that stresses the interface.

The implementation flow requires a top-level module that instantiates your design portion of the IP and the traffic generator. This top-level file will be present in the example design. The example design will have all the required constraints for the implementation of the design. The example design can be taken through the implementation flow and a bit file can be generated. You can go through the I/O planner to customize the pin out of the example design as per your PCB layout. You can also skip the I/O planner and generate a bit file with the default pin out. You can use the example design to validate the memory interface in your PCB. Memory interfaces operates at a high data rate. During system bring up, the example design flow is a good way of bringing up the memory interface in the PCB in a unit level without the other parts of the system.

The behavioral simulation of example design can be performed by selecting the *Run Simulation* option in Vivado. The Vivado simulator is supported by default and the option to support various third-party simulators is provided. The simulation waveform will be opened in the Vivado GUI framework with the relevant signals that are important for the design. The example design will have a simulation top level that instantiates the user design, traffic generator, memory model, clocking, and reset logic. The example design behavioral simulation provides you with the waveforms that show the interaction with the user interface and interaction of the memory interface signals with the controller and provides information on the latencies involved in the design.

5.3 Calibration

Calibration is a very important aspect of the memory controller design. The interface operates at very high data rates and due to that the data valid window will be very small. Good calibration techniques are required for reading and writing the data reliably for the memory interface. This section describes the concept of calibration that would be applicable for all memory interfaces.

The data is captured from the memory and written to the memory at both the edges of the clock. A *DDR4* operating at 3200 MT/s will have a clock period of 625 ps. With the dual data rate interface, the bit time would be 312 ps. Within the bit time, various uncertainties will affect the data valid window as shown in Fig. 5.1.

The uncertainties shown in Fig. 5.1 will be a combination of variations from memory, *FPGA*, and PCB. The goal of calibration is to center the capture clock in the middle of data valid window during reads and have the write clock in the middle of the data valid window during writes. For read side the following would add to the uncertainties at a high level:

- Data valid time from the memory
- Any drift with respect to clock and data from memory that is sent to the *FPGA*
- Skew between different signals that pass through the PCB
- Board inter symbol interference
- Jitter on the clock that is fed to the memory
- Setup and hold requirements of the capture flop in the *FPGA*
- Delay variations over voltage and temperature in elements used for calibration

Write side would have the following added to the uncertainties at a high level:

- Duty cycle distortion and jitter on the clock that is fed to the memory
- Package skew and clock skew from the *FPGA*
- Delay variations over voltage and temperature in elements used for calibration
- Board inter symbol interference
- Setup and hold requirements from the memory device

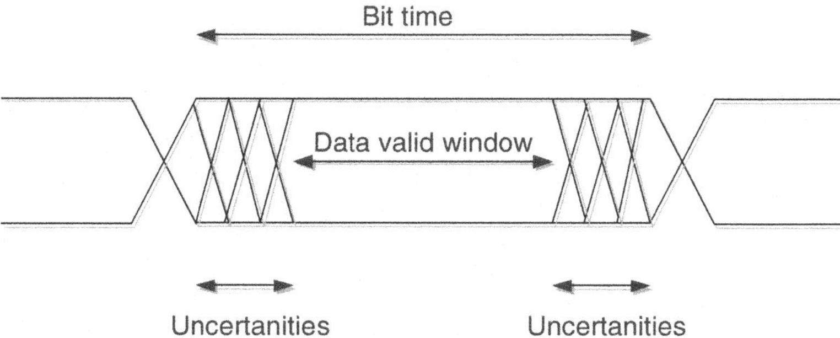

Fig. 5.1 Data bit time with uncertainties

5.3.1 De-Skew Calibration

The read and write data from the memory can be skewed based on package skew, PCB skew, clock skew, internal routing skew, and variations in the delay elements used during calibration. For the parallel memory interfaces, the data to clock ratio varies from 4:1 to 36:1. In parallel memory interface where there is one clock per multiple data bits, the skew within the data bits will affect the effective data valid window. The skew in the interface will affect the effective data valid window as shown in Fig. 5.2.

The function of the de-skew calibration would be to align all the data bits within the clock region so that the interface has the maximum data valid window for both read and write operation. The interface data valid window will be determined by the common time in which all the interfaces have valid data.

5.3.2 Read Calibration

The read clock at the output of the memory during read operations will not be center aligned with the data and will be edge aligned as shown in Fig. 5.3.

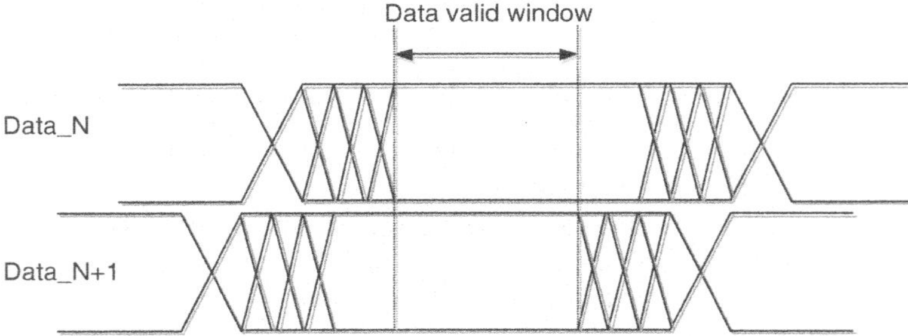

Fig. 5.2 Data bus with skew

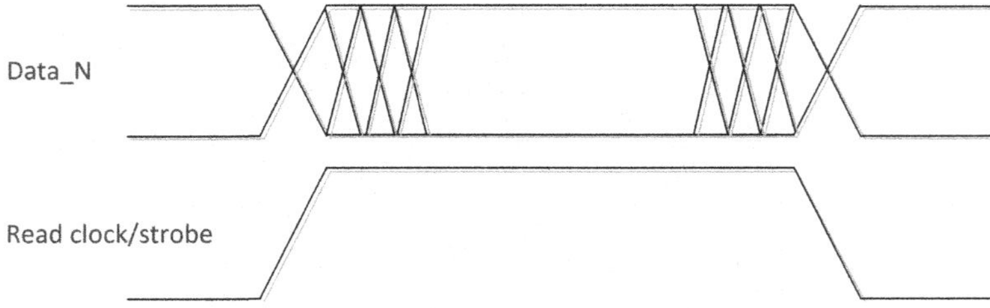

Fig. 5.3 Read data and read clock/strobe from memory

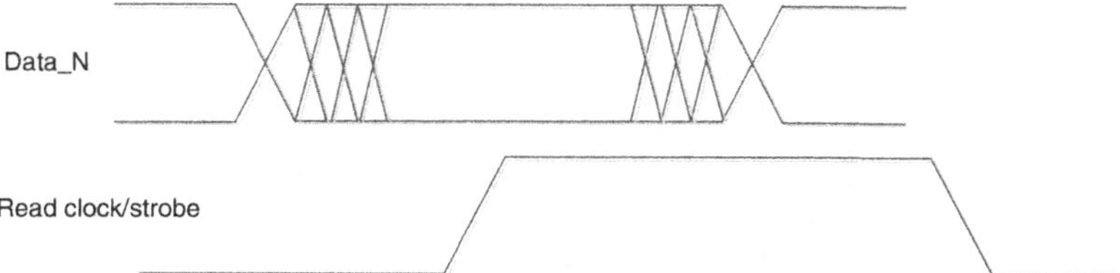

Fig. 5.4 Read data and read clock/strobe after read calibration

The main function of the read calibration is to figure out the read valid window across all the associated data bits and center the capture clock in the middle of the data valid window. This stage will usually write a pattern to memory and read from it continuously. Write calibration might not have been completed before this stage. The writes to the memory has to be successful for this stage to function properly. Some memory standards have registers within the device that has predefined data patterns that can be used during this stage and would not require a write to the memory. For devices that do not have preloaded data pattern, the read calibration will write a simple data pattern. The simple data pattern will guarantee enough setup and hold margin to make sure the writes are successful. The calibration algorithms will start with the simple pattern or preloaded patterns to complete the initial stage. After the initial calibration, for higher data rate interfaces, a complex data pattern that mimics the worst case signal integrity effects will be used to further center the clock accurately in the data valid window.

Read calibration algorithm using the training pattern will scan the data eye to figure out the uncertainty region and the effective window in which data is valid. This stage requires multiple samples of the read data to accurately figure out the uncertainty region to account for jitter and other signal integrity effects. Once the scanning is done, the calibration algorithm will position the capture clock in the center of the eye as shown in Fig. 5.4.

Read calibration in majority of the memory controller designs will include a stage for estimating the read latency. A read command will be issued by the memory controller, and it will reach the memory device after going through the delay in the address path. The memory device will have a spec for read latency, and the read data will appear on the read data bus after the read latency number of clock cycles from the time the read command was registered. The read data will have to go through the read data path delay and any other delay in the read capture stage. In most designs the read data will be captured using the clock/strobe from the memory and will be transferred to the controller clock domain for use in the other parts of the design. The memory interface will have multiple sets of data with its own read clock associated with it. All the data sets from the memory need not be aligned when it is available at the controller clock domain. Optional delay stages have to be added to align the entire interface when the data is available in the controller clock domain. The *read valid* calibration stage will estimate all the delays in clock cycles and provide the information on when the read data would be available at the controller clock domain after the read command is issued.

5.3.3 Write Calibration

The write calibration stage is required to center the write clock/strobe in the center of the write data. The memory devices have a requirement of having the write clock/strobe to be in the center during write transactions. For high-speed interfaces in which every picosecond counts, a precise calibration would be required to center the clock/strobe in the write data window. The concept behind write calibration is very similar to read calibration. The calibration algorithm would write a data pattern into memory and read it back to see if the write was successful. During the write the write clock/strobe will be moved using fine resolution delays across the data bit time to figure out the optimal position.

Write calibration in most of the controllers will have write latency calibration. Similar to read latency calibration, this stage is to calibrate out the delays that are in the write path and estimate the write latency so that the controller can satisfy the write latency requirements for the memory device. Write calibration depending on the memory technology will have an additional calibration stage to align the write clock/strobe with the memory clock. The write clock/strobe will be a point to point connection. The memory clock will go to multiple components and will have more than one load. The arrival times of the write clock/strobe and the memory clock will not be aligned and this stage is to align them both.

5.3.4 VT Compensation

VT compensation is not necessarily a calibration stage but the logic to compensate for the voltage and temperature drift that will occur over the period of time. Initial calibration will calibrate out the process variations; the dynamic variations due to VT will need compensation. There can be difference in the way variations happen between the clock path and the data path. In the worst case scenario, the data path and the clock path can drift in opposite directions. The dynamic variations can happen at any rate. The VT compensation logic would have to sense the drift and correct for it.

The compensation logic would have to monitor the drift and compensate as and when the drift happens. If left uncorrected there will be reduction in margin and in certain conditions data corruption can occur due to too much variations. The compensation logic would have to monitor the *FPGA* conditions as well as the signals from memory to detect the movement. The compensation logic would need to monitor the read data and/or read clock/strobe coming from the memory. If the user traffic does not have any read commands for a certain period of time, then the memory controller would issue read commands for maintenance purpose. The read data from these maintenance commands will not be passed on to the user interface. The interval between the maintenance commands is determined by the memory interface design requirements.

5.4 Signal Integrity

Signal integrity effects play a big part in the memory interface performance. Signal integrity simulations need to be performed for the memory interfaces, and the simulation recommendations need to be used for the PCB design as well as the memory controller options. For low-frequency interfaces the signal integrity has little effect on the signal and the interface can work reliably. At high frequencies the signal integrity effects like ringing, cross talk, ground bounce, and reflections affect the signal quality and can result in data integrity problems. Impedance mismatch is one of the key aspects that needs to be taken care of in the memory interface design. Impedance mismatch causes signals to reflect along the transmission line. The reflections can subject the signals to ringing, overshoot, and undershoot which in turn will cause signals to be sampled improperly at the receiver. The source impedance must match with the trace impedance.

Figure 5.5 shows an example of a driver, transmission line, and receiver setup. The impedance of the driver, transmission line, and the receiver have to match to avoid impedance mismatch. Various termination schemes are available for you to match the impedance. You have the option of terminating on the PCB or use the on-chip termination that is available in the *FPGA* and in the memory device. Xilinx *FPGAs* have onboard programmable termination called Digitally Controlled Impedance (*DCI*). *DCI* offers on-chip termination for receivers and drivers across multiple configurations that will satisfy your system requirement. *DCI* helps you to leave the termination implementation to the *FPGA* and simplify the PCB design. Similar to the *FPGAs*, the memory devices also have on-chip termination called On Die Termination. On the *FPGA* end, various other options are provided to improve the signal integrity. Vivado provides attributes to control drive strength and slew rate. Drive strength and slew rate can be used to tune an interface for adequate speed while not overdriving the signals. The memory wizard tool in Vivado IP catalog will automatically chose the correct setting for a given memory interface. There are certain options like *ODT* that might have multiple choices. You have the choice to go with default or chose the option that matches your requirements.

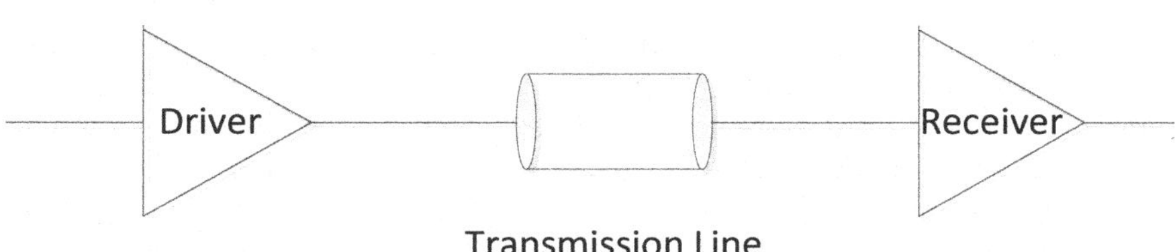

Fig. 5.5 Driver, transmission line, and receiver example

5.5 DDR4 SDRAM

DDR4 SDRAM (double data rate synchronous dynamic random access memory) introduced in 2014 is the latest (at the time of writing this book) memory standard that is widely used in the industry.

The Vivado tool provides various options for you to customize the memory controller. Based on the system requirements, you can select the options given below:

- Memory device selection: density of the device, DQS to DQ ratio, column address strobe read and write latency, memory speed grade, component, DIMM, SODIMM, RDIMM, 3DS, or LRDIMM
- Memory controller options: user interface selection, efficiency switches, addressing options for various payloads, data width, and *ODT* options
- *FPGA* options: *FPGA* banks and pins to be used, *FPGA* termination options, VREF options, and clocking options (input clock to the memory controller IP)

The most important aspect for you would be the throughput of the memory controller and the storage requirements. A 64 bit *DDR4* memory operating at 3200 MT/s will have a theoretical peak bandwidth of 25,600 MB/s. The bandwidth of the memory subsystem would depend largely on the memory configuration and the access pattern. The configuration is fixed during the initial selection. The access pattern varies based on the traffic in the system. You can take advantage of the memory controller features which will help in improving the practical bandwidth.

5.5.1 *Performance*

Efficiency of a memory controller is represented by Eq. (5.1):

$$\text{Efficiency} = \text{Number of clock cycles DQ bus was busy} \,/\, \text{Number of memory clock cycles} \tag{5.1}$$

The efficiency percentage will determine the bandwidth of the system. A 64 bit memory operating at 3200 MT/s with 80 % efficiency will have an effective bandwidth of 20,480 MB/s compared to theoretical bandwidth of 25,600 MB/s. The memory timing parameters and the memory controller architecture have a big effect on the performance of the memory subsystem. The memory timing parameters are requirements as per memory protocol, and commands can be scheduled in a way that the wait times for servicing timing parameters can be hidden or avoided.

To access a memory to perform a read or write operation, *row* access commands are required to open and close *rows* in the memory. If a *row* in the *bank* needs to be accessed, first the *row* in the *bank* has to be opened. Opening of *row* (*activate* command) has wait times associated with it to move the data from the *DRAM* cell arrays to the sense amplifiers and having it ready for read or write operations. To close a *row* in a *bank*, a *precharge* command has to be issued. The *precharge* command has

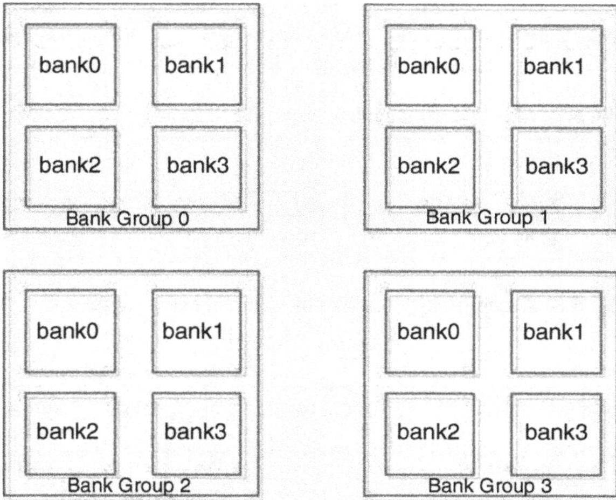

Fig. 5.6 Bank grouping for four bank group configuration

its own timing requirements to reset the sense amplifiers and get it ready for another *row* access command. At a given time only, one *row* can be kept open in a *bank* of memory. *DDR4* memory has 16 *banks* and at any given time one *row* in each of the 16 *banks* can be kept open.

To get the required performance, the number of *row* access commands has to be minimized, and more of *column* access commands (read or write) have to be issued. *DDR4* memories also have a concept called *bank groups*. Each *bank group* will have four *banks* associated with it as shown in Fig. 5.6.

The new feature in *DDR4* is that access across *bank groups* has less access time compared to access within the *bank group*. In terms of *bank* access, the example shown in Fig. 5.7 has less access time which helps in performance. The example shows write commands; the same is true for read commands as well. The memory controller will be able to keep multiple *banks* open and can hide the *row* access times between the column commands. The *burst length* for *DDR4* is eight, and due to the dual data rate for every four memory clock cycles, there will be eight data transfers. Back to back *column* commands can be issued only every four memory clock cycles. Between the *column* commands, the *row* commands can be interleaved to open and close *banks*. *Column* access that changes *bank groups* every four clock cycles will have the advantage of minimum access time. Access across the *bank groups* in most of the scenarios will avoid idle cycles between the *column* accesses.

The other access pattern that can hide the *row* access penalties to a certain extent is shown in Fig. 5.8. The performance of this access pattern will not be as good as the performance of access pattern shown in Fig. 5.7. By switching to different *banks* within the *bank groups* or across the *bank groups* gives the flexibility for controller to have multiple *banks* open and schedule commands in such a way that the *row* access times can be hidden. Switching of *banks* within the *bank groups* is not guar-

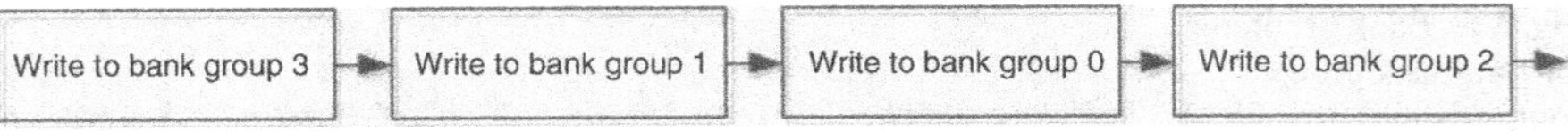

Fig. 5.7 Access across bank groups

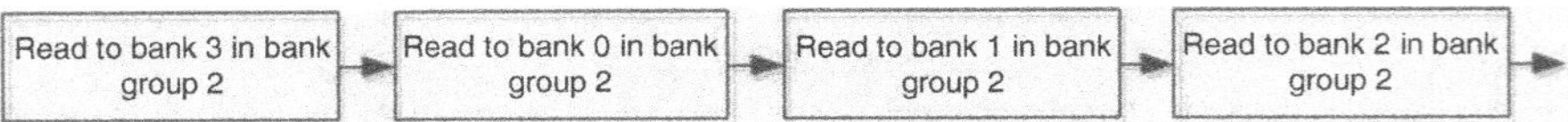

Fig. 5.8 Access to different banks within a bank group

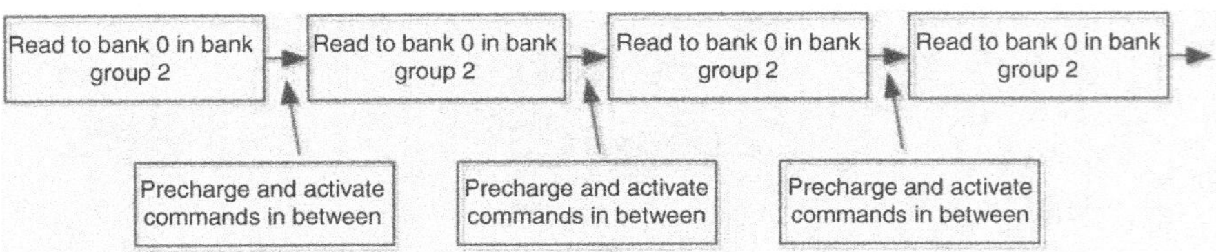

Fig. 5.9 Access to different rows within a bank

anteed to avoid the idle cycles between *column* accesses as the access times are higher in this scenario compared to accesses across *bank groups*.

In terms of performance, random access pattern will incur more *row* access penalties. An example of a pattern that will have low performance is shown in Fig. 5.9. In the example shown in Fig. 5.9, the read commands go to different *row* addresses in *bank* 0 that is present in *bank group* 2. The accesses are within the *bank*, and every time a different *row* is opened, the existing *row* in the *bank* has to be closed and the new one opened. The controller has to wait for the closing and opening times before issuing the *column* commands. The timing requirements for the *activate* and *precharge* commands will be more than the time that is between the two *column* commands. This results in idle cycles between *column* accesses.

The data bus for *DDR4* is a bidirectional bus. Every time there is a change from write to read or read to write, it takes time to reverse the direction of the data bus. Most controllers have reordering functionality built in them to group reads and writes to minimize the occurrence of turnaround time.

The memory controller solutions from Xilinx provide options for you to map the user address to the address bus of the *SDRAM*. Depending upon the option selected by you, the user interface in the controller maps the address from the user to the *SDRAM* address bus. Based on your selection, the parts of the address bits would be assigned to *rank*, *bank group*, *column*, and *row* bits of the *SDRAM*. The mapping can have an impact on the memory controller performance. The controller would be able to make use of the controller resources and able to keep the data bus busy with *column* commands.

There are also memory maintenance commands like *refresh* and *ZQCS* that have to be issued to the *SDRAM* periodically. The memory controller by default will issue the maintenance commands periodically to satisfy the *SDRAM* requirements. When the memory controller issues these maintenance commands, a long *burst* might be broken up affecting the efficiency. The maintenance commands also have the requirement to close all the open *banks*, and they have to be opened again after the completion of the maintenance commands. The user can choose to take control of the maintenance commands and issue it through the user interface signals to improve efficiency. Care should be taken to make sure the *SDRAM* timing requirements are met when the user takes over the responsibility to issue the maintenance commands.

DDR4 has wide use in many applications. *DDR4* comes in various form factors to suit the different system requirements. The common use of the memory is in desktop, laptop, and servers as the main system memory. *DDR4* is highly suited for processor-based systems and in any application that require mass storage. *LPDDR4* memory interface has similar features like *DDR4* with additional low-power features. *LPPDR4* memory is not discussed separately and majority of the concepts described in *DDR4* are applicable to it.

5.6 RLDRAM3

RLDRAM3 (reduced latency dynamic random access memory), introduced in 2012, is the latest offering from Micron Technology, Inc., on the reduced latency *DRAM* category. *RLDRAM3* has the advantage of reduced latency combined with good storage capacity. Similar to *DDR4* the Vivado tool provides various options for you to configure the memory controller.

5.6.1 Performance

Efficiency equation mentioned in Eq. (5.1) (Sect. 5.5.1) is applicable to *RLDRAM3* as well. The memory timing parameters and the memory controller architecture have a big effect on the performance of the memory subsystem. *RLDRAM3* has two important timing parameters that affect performance: *tRC* and *tWTR*. *tRC* (*row* cycle time) is defined as "after a read, write, or *auto refresh* command is issued to a *bank*, a subsequent read, write, or *auto refresh* cannot be issued to the same *bank* until *tRC* has passed." *tWTR* (write to read to same address) is defined as "write command issued to an address in a *bank*; a subsequent read command to the same address in the *bank* cannot be issued until *tWTR* has passed."

RLDRAM3 has 16 *banks*. As shown in Fig. 5.10, if the access is scheduled in such a way that the same *bank* to *bank* access comes in after *tRC* time requirement,

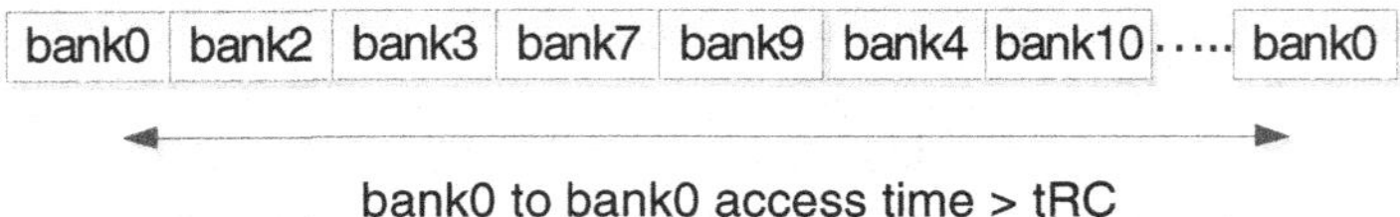

Fig. 5.10 Access across banks

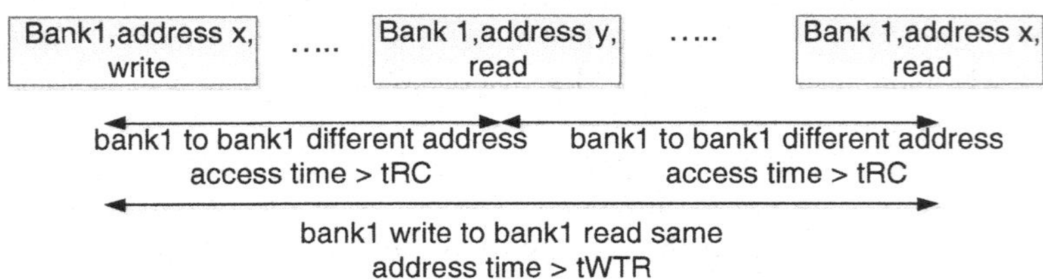

Fig. 5.11 Bank access with tRC and tWTR

then the efficiency will be high. There will not be any idle cycles between the read and write commands. If the traffic pattern is such that the tRC time is not satisfied, then the controller has to pause so that *tRC* time can be elapsed before issuing the command. The read to write and write to read turnaround as per specification is one memory clock cycle. Whenever there is a turnaround requirement, the controller has to pause for one memory clock cycle. Some memory controllers due to clock ratios and I/O requirements might end up waiting for more than one clock cycle.

Write followed by *read* to the same address in the *bank* will have a larger wait time. The tWTR parameter comes into effect when a write is followed by read to the same address in the *bank*. The controller would have to pause the traffic and wait for tWTR to elapse in this scenario which will have an effect on efficiency. As shown in Fig. 5.11, the write to read to the same address in the *bank* has to be spaced apart to satisfy the tWTR requirements. Figure 5.11 also shows an example of write to read to different addresses within a *bank*. In this scenario only the turnaround time and the tRC requirement will come into effect. The same is true for read to write within a *bank* for any address; the turnaround time and the tRC requirement will come into effect.

The low-latency and high-bandwidth characteristics of *RLDRAM-3* are highly suited for high-bandwidth networking, L3 cache, high-end commercial graphics, and other applications that require the *RLDRAM3* features.

5.7 QDRIV

QDRIV SRAM (quad data rate IV synchronous random access memory) introduced in 2014 is the latest offering from Cypress Semiconductor on the synchronous *SRAM* category. *QDRIV* has lower latency and does not have any timing parameters that affect efficiency. The Vivado tool provides various options for you to customize the memory controller similar to *DDR4*.

QDRIV memory device has two independent bidirectional data ports. Both the ports operate at *DDR* data rate and can be used for both read and write transactions. One common *DDR* address bus is used to access both the ports; rising edge is used for accessing one port and the falling edge for the other port. The ports are named port *A* and port *B*. Each port has its independent read and write clocks. Port *A* address will be sampled at the rising edge of the address clock, and the port *B* address will be sampled on the falling edge of the address clock.

There are two types of *QDRIV* parts: *XP* and *HP*. *HP* parts do not have any restriction on the access between two ports. XP parts have some restrictions and the *bank* access rules are listed below:

- Port *A* can have any address on rising edge of the address clock. There is no restriction for port *A*.
- Port *B* can access any other *bank* address on the falling edge of the clock other than the *bank* address used by port *A* on the rising edge.
- Port *B* can access any address in the falling edge if there was address presented on rising edge for port *A*.
- From the rising edge of the input clock cycle to the next rising edge of the input clock, there is no address restriction.

The most important aspect for you would be the throughput of the memory controller and the storage requirements. A 36 bit *QDRIV* memory operating at 1066 MHz will have a theoretical peak bandwidth of 153.3 GB/s. The bandwidth of the memory subsystem would depend largely on the memory configuration and the access pattern. The configuration is fixed during the initial selection. The access pattern varies based on the traffic in the system. *QDRIV* interface does not have any timing parameter that affects the performance. It has only one restriction on *bank* access between the two ports. If the memory is accessed in a way that takes advantage of the *QDRIV* features, then 100 % bandwidth can be achieved which is not possible in other memory technologies.

5.7.1 Performance

Efficiency equation mentioned in Eq. 5.1 (Sect. 5.5.1) is applicable to *QDRIV* as well. The access pattern of the user will have an effect on performance for XP *QDRIV* devices. In a given clock cycle, port *B* cannot access the same *bank* address as the *bank* address used by port *A*. If the traffic pattern is such that there is banking violation in port *B*, then the memory controller would have to pause the traffic to take care of the banking restriction. Other than that the only time there will be an effect on efficiency would be the *read* to *write* and *write* to *read* turnaround times. The user would have to make sure to group the reads and writes to get maximum efficiency. Since port *A* and port *B* have independent data buses, there is no

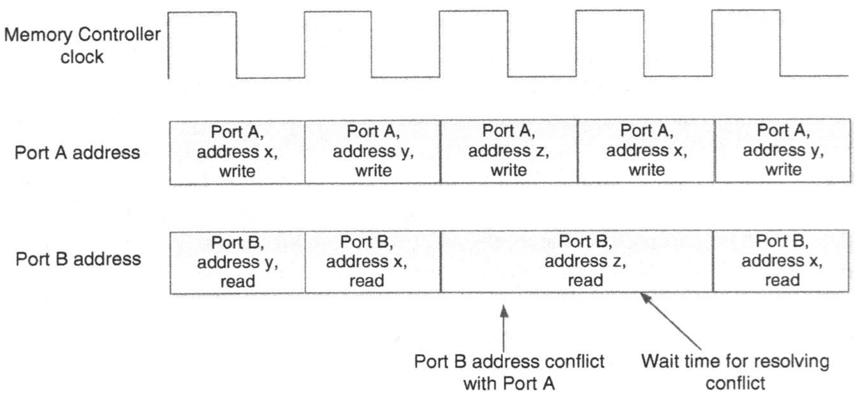

Fig. 5.12 Port A and Port B access with conflict

restriction on *read* and *write* between the ports. The turnaround wait time is within the port data bus.

Figure 5.12 shows an example of conflict when port *B* accesses the same *bank* address as port *A*. On every memory controller clock cycle, the commands for port *A* and port *B* can be accepted. The corresponding port commands will be sent on the rising and falling edge of the memory clock to the *QDRIV SRAM* device by the memory controller. Usually the memory controller will be operated at a lower clock frequency than the memory interface frequency for timing reasons. In this example it is shown that the memory controller operates at the memory interface frequency. The controller would have to stall certain number of clock cycles to resolve the conflict. In *QDRIV* case the stalling would be only one clock cycle unless other factors come into play.

QDRIV memory is attractive for applications that would require high efficiency for random traffic. Latency would also be critical for those applications and *QDRIV* provides low latency at higher data rates. Typical applications that would use *QDRIV* are high-speed networking, communication, and any application that would have access that would be random in nature.

Chapter 6
Processor Options

Siddharth Rele

6.1 Introduction

FPGAs are great for high performance and parallel computing. Addition of a processor enables control path processing which is required for most applications. Xilinx FPGAs allow you to make use of processors, which could be soft (implemented on fabric), or hard (pre-built). Designing with processors on FPGA has been made easier through use of Xilinx Vivado IP Integrator and SDK tools. This chapter will explore the usage of both hard and soft processors within Xilinx FPGAs for some typical applications.

6.2 Computing on FPGAs

A basic introduction to FPGAs has been provided in Chap. 1. Processors and FPGAs provide similar general processing capabilities but are differentiated by the way you use them for programming and the type of applications/use cases. Processors use software methods, while FPGAs were traditionally programmed through use of hardware design languages such as Verilog and VHDL. Processors are good when it comes to control flow as well as processing based on control, while FPGAs are preferred when unconditional processing has to be done on a larger data-set.

There are several benefits of combining the control capabilities of a processor with data-intensive compute of the FPGA. Solutions can be developed by having a

S. Rele (✉)
Xilinx India Technology Services, Hyderabad, Telangana, India
e-mail: siddharth.rele@xilinx.com

S. Churiwala (ed.), *Designing with Xilinx® FPGAs*,
DOI 10.1007/978-3-319-42438-5_6

multi-chip solution with standard discrete processors connecting to FPGAs over I/O pins. While these solutions work, they lead to increased latencies which may not be acceptable for several high-speed applications. Hence, an integrated processor with FPGA solution would be more apt for most applications. These lead to lower latencies when processor accesses rest of the design as well as leads to reduction in overall I/O count.

6.3 Processors on FPGAs

There are two categories of processor solutions possible on Xilinx FPGAs:

- Soft processors
- Hard processors

You can choose either of these depending on the amount of control processing and I/O required for your application. The cost of the overall solution will also play a role in making the decision.

6.3.1 Soft Processors

Soft processors make use of the FPGA fabric for implementing the processors. For low speed (200 MHz and below), soft processors are a good option. These come in multiple flavors and are user configurable. Depending on the nature of the application, you can choose to trim down the functionality from the processors.

At the lowest end, you can implement an 8-bit processor with bare minimal instruction set. One example of such a processor is PicoBlaze™ from Xilinx. This processor is a good replacement for state machines. PicoBlaze does not have a compiler tool chain and hence requires the program to be written in assembly language. This program is stored in the local memory available on the FPGA as a memory store. The simplicity of the architecture enables a processor which can be implemented in just about 26 slices.

As a step up, Xilinx introduced a 32-bit highly configurable processor named MicroBlaze™ in early 2000. This RISC-based soft processor is capable of achieving clock speed of around 400 MHz on the UltraScale FPGA architectures. It supports an option of a three-stage or a five-stage pipeline, configurable cache, optional multiply and divide units, optional barrel shifter, single- and double-precision floating-point unit, and more. Every additional feature selected in hardware will lead to usage of FPGA resources and can have impact on the max frequency (F_{max}) possible. The choice can be made based on the needs of the user application. For example, if there are many multiply operations to be done, it is better to enable a hard multiplier. It can save over 1000 clock cycles for every multiply operation done over a software library-based solution (Fig. 6.1).

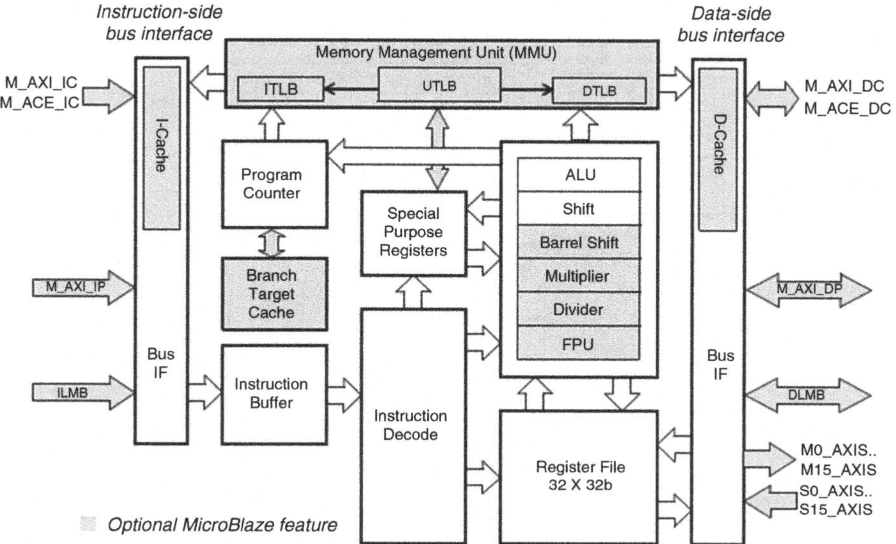

Fig. 6.1 MicroBlaze processor

In addition to the soft processors from Xilinx, you can also build your own processors or procure one from IP vendors and open source. There have been implementations of ARM done on Xilinx FPGAs in academia as well as industry. There have also been implementations of processors with reduced instruction set targeted toward specific applications. MicroBlaze has been optimized for FPGA implementation and is usually better suited for both resource count as well as $F_{\max}$.

6.3.2 *Hard Processors*

While soft processor can cater to the needs of mid-level applications, there are a few factors that make hardened processors on FPGAs a key requirement. Some applications require high-speed processing of 1 GHz and above. There are several software applications which are targeted toward standard processors like ARM. Retargeting these to other processors specific to FPGAs could take up a lot of effort and reverification (Fig. 6.2).

Xilinx introduced Zynq-7000™ family of devices which includes a complete SoC with two Cortex-A9™ processors along with a configurable mix of peripherals. These include high-speed peripherals such as *GigE* and *USB* and low-speed peripherals like *SPI*, *I2C*, *CAN*, and *UART*. The processing system (PS) also includes controllers for various volatile memories (*DDR3*, *DDR2*, *DDR3L*) as well as flash memories (*QSPI*, *NOR* and *NAND*). By hardening the most commonly used blocks in the SoC, Xilinx has enabled saving FPGA logic for the key acceleration logic rather than using it for interfacing to components on the board.

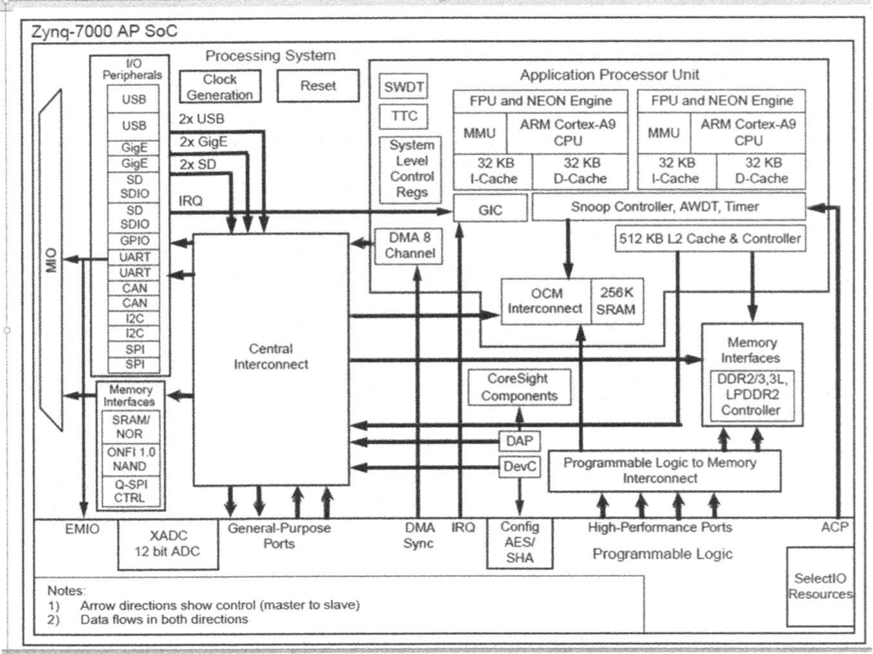

Fig. 6.2 Zynq-7000 block diagram (not to scale)

The *PS* is built such that the SoC can be used even without the programmable logic (*PL*) fabric turned on. This enables the software users to be productive even without the FPGA design has been created. Section 6.5 (Putting It All Together) talks about some of the ways to use Zynq-like devices.

The architecture was further extended in the Zynq *UltraScale+ MPSoC*™ shown in Fig. 6.3. Xilinx raised the compute power of the SoC by introducing four Cortex A-53™ cores and two Cortex-R5™ cores. In addition to these processors, there is a *GPU* as part of the SoC as well. With MPSoC, Xilinx has also introduced a host of high-speed peripherals which include *SATA*, *DisplayPort*, *PCIe*, and *USB 3.0*. These are built on top of high-speed SerDes which are part of the SoC. Xilinx extended the memory support to *DDR4* as well. Security and isolated power domains have been two major advancements in Zynq UltraScale+ MPSoC. The processor and other masters in the SoC can have secure access to specific peripheral/memory through the Xilinx Peripheral Protection Units (XPPUs) and Xilinx Memory Protection Units (XMPUs). Since the SoC packs a lot of powerful peripherals, the power consumption has to be controlled. Xilinx has split the SoC in a lower power domain and full power domain making it easier for customers to split their application appropriately and shut down peripherals when not in use.

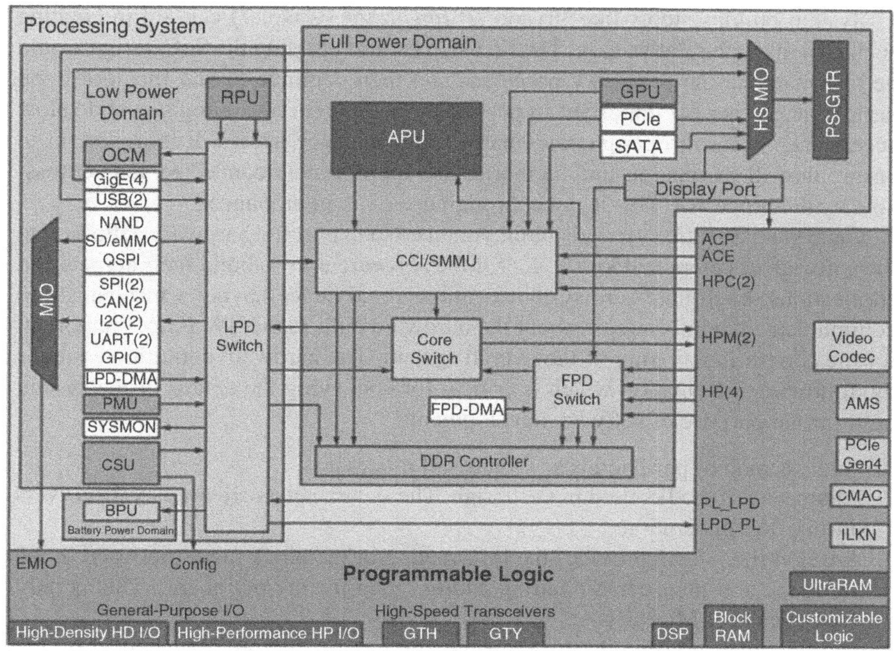

Fig. 6.3 Zynq UltraScale+ MPSoC

6.4 Tool Chains

While processor architectures are a key factor for designs, it is equally important to have appropriate tools in order to build systems which can integrate the programmability of FPGAs and processors. These tools include hardware designs tools at a higher level of abstraction as well as traditional software design and debug tools.

6.4.1 Integration Tools in Vivado

A completely functional SoC consists of a processor, soft peripherals, as well as memories. The first step toward building an SoC+FPGA system is to identify the processor of your choice (primarily MicroBlaze or a Zynq class of SoC). The next step is to determine the correct memory architecture suitable for your application. This includes memories internal to the FPGAs (such as *block RAMs*) and external memories (Chap. 5) which range from nonvolatile flash memories to volatile SRAM and SDRAM memories.

System building tools like Vivado *IP Integrator* (Chap. 7) can aid in creating such designs with relative ease. The first step toward building the SoC design would be to instantiate the necessary processor and then continue to add memories and peripherals. It is also important to partition your system into a register-style slow access (i.e., access to peripherals) and a faster memory access. It is important to ensure that all peripherals and memories are appropriately connected and addressable by the processor. This is done through use of an interconnect.

Once the hardware system is built, you need to export the hardware definition to the software development kit (SDK) for the software user to build their kernels and applications. Each hardware system is unique, and hence having a mechanism to communicate the information about the hardware built to the SW *BSP* is important for the correct functioning of the overall system. The hardware definition contains the following information which is critical for conveying the details of the system built for the purpose of software development:

1. Address map of peripherals as seen by all processors.
2. Parameters of the IPs used in the design. These parameters are used by the drivers during BSP generation.
3. A BMM file, which provides the information of the block memories used in the hardware and their corresponding address map for the peripheral. This is only used in case of MicroBlaze.
4. The bitstream which is built in Vivado and can be directly downloaded by the SDK during software development.

All this information is critical for the software BSP to be created. Any changes in the hardware require a reexport of the HW information and a regeneration/recompile of the SW BSP.

6.4.2 *Compilers and Debuggers*

Embedded application developers typically write their software programs in a higher-level language like *C* and *C++*. These high-level programs are translated into assembly-level object code and eventually into an executable and linking format (ELF) file which contains machine-level code. Compilers play an active role in optimizing the generated code using the context of the processor being used. For example, in case of MicroBlaze, if the multiplier is implemented as part of the SoC, the code generated would use the *mul* instruction. If it is not, it would make a call to the *multiply* function from the pre-built libraries.

Some SoCs such as the MPSoC have more than one processor. These processors can be made to work as *SMP* (symmetric multiprocessing) or *AMP* (asymmetric multiprocessing). In case of an SMP system, the software kernel such as Linux will take care of scheduling processes on appropriate processor based on system load. With industry standard processors (ARM A9, Cortex A-53, and R5s) on the SoC, you can find the right software kernels for their systems which can be used as a base. With AMP system, you need to take care of not just execution on the individual

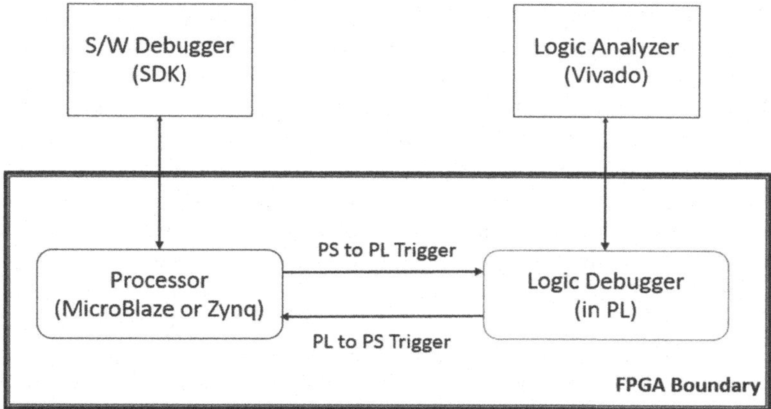

Fig 6.4 Cross trigger (conceptual diagram)

processors but also have the appropriate communication mechanism between the various processors.

Often, programs need to be debugged in order to get to the correct functionality. Software tool chains such as SDK would be incomplete without a debugger which can connect to the board and provide helpful information about the program execution.

There are certain situations where it is important to debug the software processes running on the processor (*PS*) in conjunction with the hardware transactions in the fabric (*PL*). Xilinx supports *cross trigger* solution for both soft processors (MicroBlaze) and hard processors (Zynq-7000 and Zynq UltraScale+ MPSoC). A conceptual diagram is represented in Fig. 6.4.

When a breakpoint is hit in the software debugger, it raises a *PS* to *PL* trigger. This trigger can be connected to a logic debug IP (such as "Integrated Logic Analyzer" [ILA] IP). The logic analyzer tool which is part of Vivado will then be used to display the current transactions on the signals being tapped in the hardware. In a similar fashion, you can generate a trigger from the *ILA* (i.e., *PL*) to the *PS*. This will stop the processor from executing instructions leading to a breakpoint. Chapter 17 explains the triggers in hardware.

The cross trigger capability can be extended to multiple processors and multiple hardware triggers in the *PL*. It can be an extremely useful way of debugging HW/SW designs.

6.4.3 Device Drivers, Libraries, Kernels, and Board Support Packages

For using peripherals in software, it is important to know the exact function and the register map of the peripheral. Peripheral developers would typically provide a device driver which provides the APIs for accessing information at a high level.

Software applications are not all written from scratch. Many applications are built using pre-built libraries or libraries obtained from third parties. One good example of this would be the *Light weight IP* (LwIP) stack. This provides the basic Ethernet packet header processing capabilities. Applications can use the high-level APIs provided by the library in order to get their job done.

Most applications are written on top of an operating system (also known as kernels). Linux is the choice for several embedded applications. A device tree is used to communicate the details of the hardware with the Linux kernel. This includes the type and width of the devices, interrupt ids, and their addresses.

All the software components above are put together in a bundle which is called as a *board support package* (BSP). BSPs are typically tied to a specific board/SoC and the hardware for which the application is expected to be written. Once the hardware is finalized, the BSP would rarely change. The BSPs would also have standard APIs, and hence the software developers are free to write their code according to their requirements and not worry about the basic device accesses.

6.4.4 Beyond Traditional System Design

FPGA and SoC combination is now helping go beyond the traditional SoC market and providing useful acceleration techniques. Xilinx is now adding support toward software-defined flows. These flows enable offloading of software applications on hardware through underlying usage of high-level synthesis (HLS) tools. This enables embedded software application developers to off-load a compute-intensive complex algorithm to the fabric.

6.5 Putting It All Together

The best use of FPGAs and processors can be explained through a couple of simple applications.

6.5.1 Basic Application

One of the basic usages of soft processors is to act as a microcontroller which monitors a video pipeline engine and responds to interrupts when something unusual is observed in the video pattern. For example, a security camera can detect moving patterns and compare faces to a central database. On finding a match with a person of criminal background, it can set an alarm which can notify the right officials of the presence of such a person on premises. A system can be represented in a block diagram as shown in Fig. 6.4. The block diagram shown in Fig. 6.4 is overly simplified to explain the concept and the role of the processor. In a real system, there would be an additional requirement of flow control as well as more connections.

Xilinx provides IPs such as HDMI controllers, memory controllers, interrupt controllers, and MicroBlaze for building systems, while you will have to provide the special secret sauce such as facial recognition and the database lookup and compare code. The solid arrows show the typical dataflow from the external camera image and the memory lookup, while the dashed lines indicate the control flow. The MicroBlaze processor controls all the IPs in the system and is usually responsible for initialization and periodic status checks. This will depend on the software developed. Without a processor, you would have to write up complex state machines to ensure that the entire system works in tandem.

6.5.2 Advanced Applications and Acceleration

You could potentially decide to port the entire application to an SoC family of devices. The MPSoC has all the peripheral IPs necessary for realizing a system as shown in Fig. 6.5. This would require the secret sauce (such as the facial recognition and database lookup) to be written in a software, which is compiled to the ARM processor. But even with a 1.5 GHz processor, it is hard to match the performance of dedicated computation in an FPGA.

The facial recognition and comparison algorithms could take thousands of clock cycles to detect an image and compare it to a picture in the database. If the job is done in a hardware (i.e., programmable logic), the entire algorithm could be parallelized and be done in a few clock cycles. Xilinx tools make it easy for embedded algorithm developers to take the compute-intensive functions through *HLS* tool

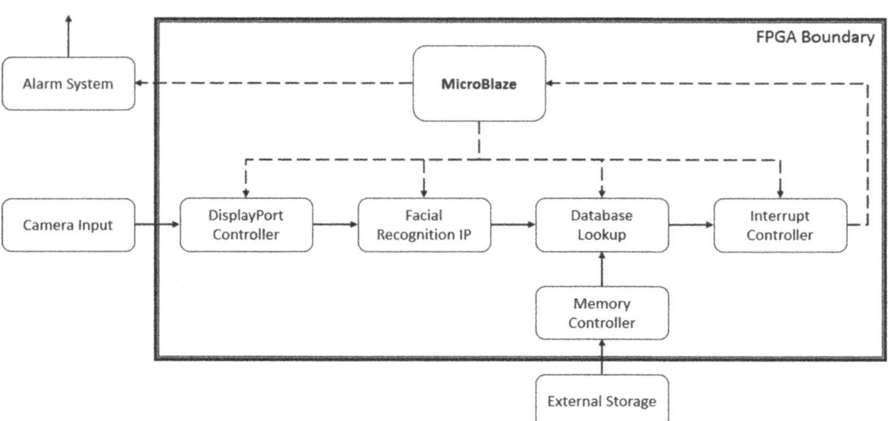

Fig. 6.5 Basic application

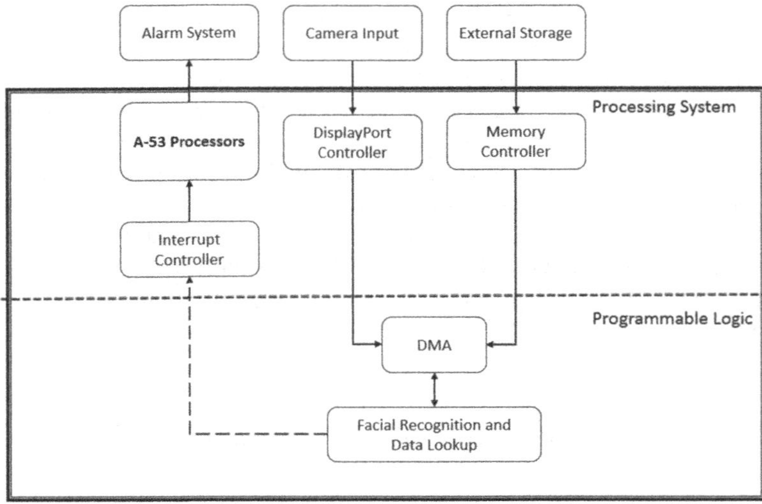

Fig. 6.6 Accelerated system with MPSoC

chain to create a hardware component which is a faster and a parallel version of the software. The tool chain also creates the necessary connection with the processor which can continue to take care of data acquisition and overall control. A simplified view of the application after running through the tool chain would look similar to the one shown in Fig. 6.6.

Chapter 7
Vivado IP Integrator

Sagar Raghunandan Gosavi

7.1 Introduction

Vivado IDE provides the **IP integrator** (IPI) with graphic connectivity canvas to select peripheral IPs, configure the hardware settings, and stitch together the IP blocks to create the digital system. Since IPI makes very heavy usage of IPs, it would be good to have a good understanding of Vivado IP Flows (explained in Chap. 3), in order to get a full appreciation of workings under the hood as you use IPI.

IPI offers many useful features that enable you to graphically design their system. The following are the main features that ensure ease of complex design creation:

- Graphical interface called block design (canvas) to construct complex IP-based designs
- TCL-based complete support for design creation
- Support for auto-connection of key IP interfaces
- One-click IP subsystem generation
- Design rule checks
- Parameter propagation
- Address map generation

After going through this chapter, you will have a deeper understanding of the features offered by the *IP Integrator* flow within the Vivado Design Suite and you will be comfortable in designing and constructing your own systems using the same. This will include instantiating individual IP components, making necessary connections between the interfaces, defining and connecting clocks and resets, configuring

S.R. Gosavi (✉)
Xilinx, Hyderabad, Telangana, India
e-mail: gosavi.sagar@gmail.com

S. Churiwala (ed.), *Designing with Xilinx® FPGAs*,
DOI 10.1007/978-3-319-42438-5_7

the settings of IP, defining the address range of slaves for their respective masters (useful in case of processor-based systems), understanding the parameter propagation, and thus generating the output products of their system.

7.1.1 Design Reuse

With the complexity of the systems increasing exponentially, it becomes extremely important to be able to reuse designs with minor modifications to their systems without the need to completely redo the design. IPI provides the right means to achieve this wherein it empowers you to configure the individual components as per the requirement in your design to construct the different flavors required for the system. It offers the ability to package your design which can be reused in other projects.

7.2 Terminology

7.2.1 Block Design (BD)

Vivado IDE provides the capability to create a workspace for you wherein you can graphically create design systems in an IPI-provided canvas, stitch the design together using the automation tools, ensure the correctness of the design, and generate the design. The block design can be created in both project and non-project mode (explained in Chap. 2). As stated above, one of the major features of the block designs is the graphical canvas which allows you to instantiate IP blocks from the *IP Catalog* and construct designs. Figure 7.1 shows the block design creation and the canvas of the BD.

7.2.2 Automation Notifications

One of the key aspects of IPI is the provision of the connection and board automation. Whenever IPI identifies potential interface connections between various IP blocks, it notifies you about the possible availability through a hyperlink visible at the top of the canvas, as shown in Fig. 7.2. For example, clock, reset, and AXI connections between the AXI-compliant IPs are covered in this automation. Detailed explanation is covered under Sect. 7.3.2 (Designer Assistance).

7.2.3 Hierarchical IP

IPI provides a feature where an IP can pack another block design within itself, thus offering another level of block design inside top level to display the logical configuration of the parent. These hierarchical blocks enable you to view the contents of the

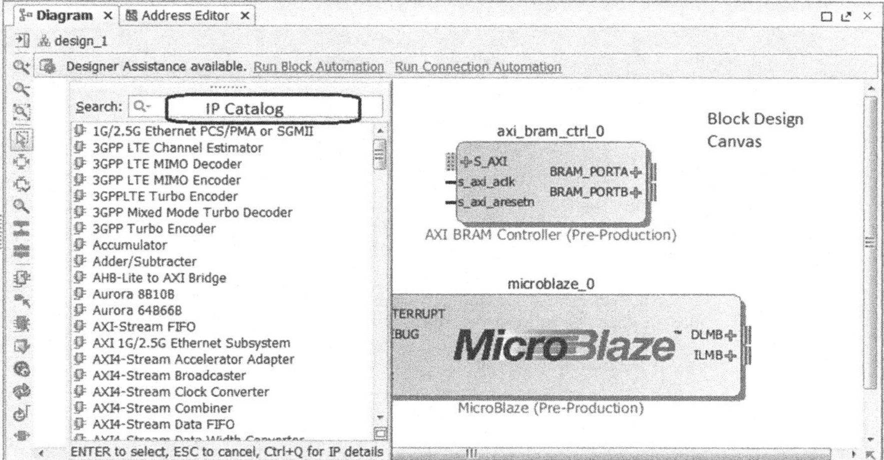

Fig. 7.1 BD canvas of IPI

Fig. 7.2 IPI notifying about automation availability

block but do not allow to edit the hierarchy. Changes are permitted only to the top level exposed parameters available in the configuration window of the IP.

7.2.4 Packaging

IPI also provides a feature wherein you can package the entire block design after it has been validated and functionality has been proven. This allows you to reuse the IP block design in other projects as well. Figure 7.3 depicts the selection window for packaging the project.

Once the block design is packaged, the tool copies the necessary files in the specified directory and adds the IP repository to the project locally. The properties associated with the package can be changed while packaging the design, thus enabling you to use the block design in other projects.

7.3 IPI Flow

7.3.1 Design Entry Within BD Canvas

The basic method of design entry in a project mode within IPI relies on instantiating the IPs from the *IP Catalog* in the block design canvas. Section 3.2 explains about *IP Catalog*. While creating a design, you need to just drag and drop the IP from the

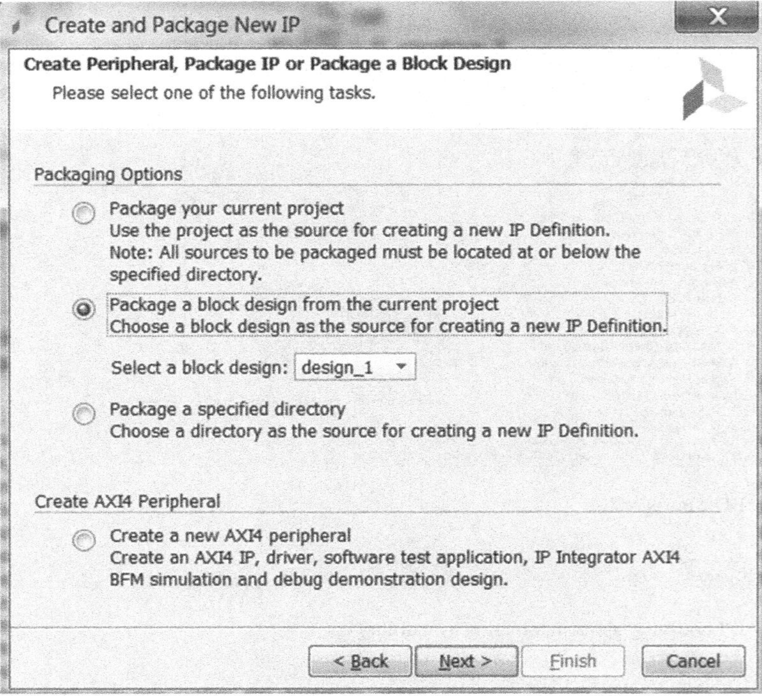

Fig. 7.3 Selection window for packaging

catalog in the canvas or can directly add to the canvas by clicking the "+" button. The IPs instantiated in the design can be individually configured based on the design requirement provided that the IP under work has those options available while it was being packaged:

- Stitching the design

  Various blocks of IP modules instantiated within the block design canvas can be respectively connected to structure the system. Block design by default automatically identifies the AXI interconnect interfaces, clock, and reset connections. This assists the users in stitching the design together.
- Ports

  Create Port option within IPI provides you with more control in specifying the input and output, the bit-width, and the type (such as clk, reset, and data). With this option you can also specify the frequency for clock ports. There is also a provision for making the port from the IP external meaning that it would be promoted to the top level.

7.3.2 Designer Assistance

Another powerful feature offered by IP integrator is the *Designer Assistance* which includes block automation and connection automation. To narrate it in brief, this feature provides users with suggestions to establish potential connections between interfaces of the compliant IPs:

- Connection automation
  This feature assists the users in connecting the AXI interfaces, reset/clock ports, and/or ports of the IPs to external I/O ports. These ports if made external will appear in the top-level HDL wrapper, and an appropriate XDC constraint would be required to be defined for them.
- Block automation
  This feature is available only when an embedded processor such as the Zynq 7000 Processing System or Zynq MPSoC or MicroBlaze processor or some other hierarchical IP such as an Ethernet is instantiated in the *IP Integrator* block design. This feature allows users to configure the related IPs based on their design requirements. It comes with a certain set of options, which you can choose from to configure the IP, thus bypassing the need to manually configure the IP.
  For example, in Fig. 7.4, once the MicroBlaze processor IP is instantiated in the design, the block automation becomes available.
  On clicking the *Run Block Automation*, a pop-up shows up as shown in Fig. 7.5, which allows you to configure the MicroBlaze IP.
  Once configured, the block design updates to reflect the changes selected, and a new set of IPs also appear in block design based on the set of selection.

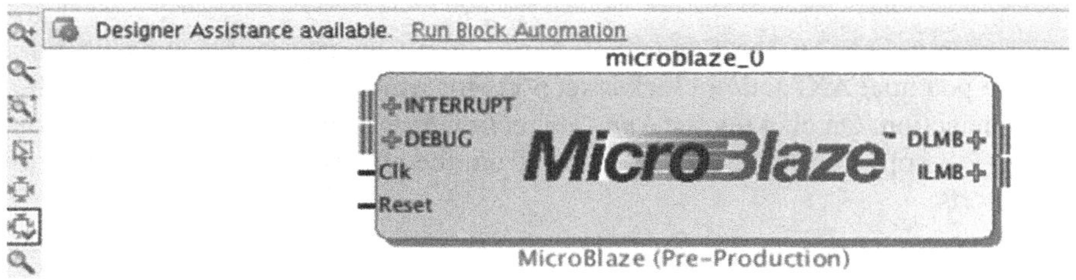

Fig. 7.4 Block automation notification

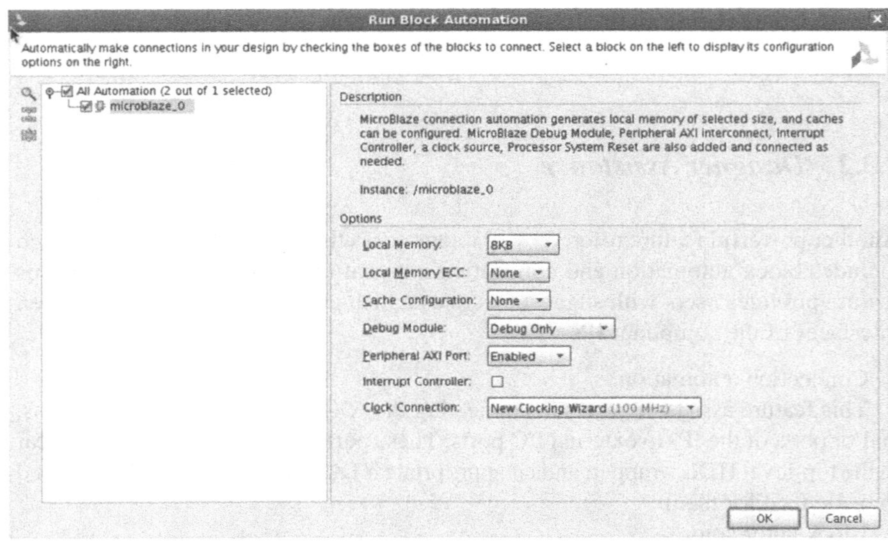

Fig. 7.5 Block automation configuration settings for MicroBlaze

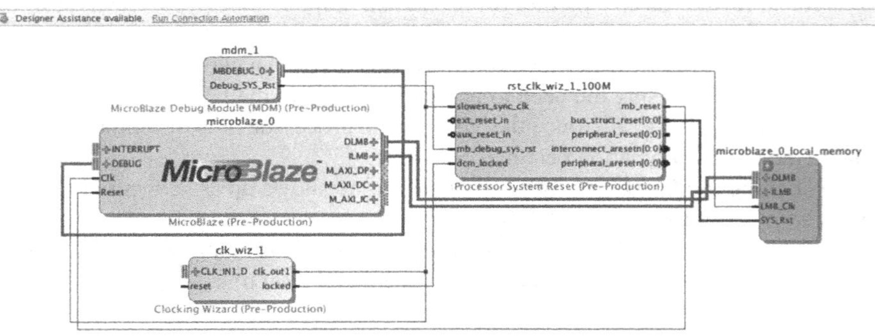

Fig. 7.6 Block design after execution of block automation

As seen in Fig. 7.6, the *Connection Automation* gets activated as it has identified the potential AXI and/or clock/reset ports for which it can assist you to establish connection. On clicking the *Run Connection Automation*, a window as shown in Fig. 7.7 pops up. You can then choose from the available set of selections for these ports.

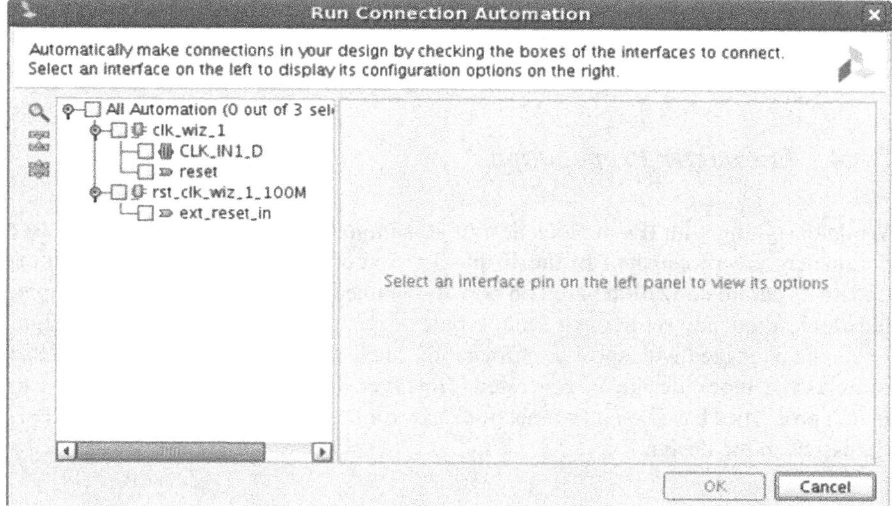

Fig. 7.7 Configuration settings for connection automation

Cell	Slave Interface	Base Name	Offset Address	Range	High Address
microblaze_0					
Data (32 address bits : 4G)					
microblaze_0_local_memory/dlmb_br...	SLMB	Mem	0x0000_0000	8K	0x0000_1FFF
axi_bram_ctrl_0	S_AXI	Mem0	0xC000_0000	8K	0xC000_1FFF
Instruction (32 address bits : 4G)					
microblaze_0_local_memory/ilmb_bra...	SLMB	Mem	0x0000_0000	8K	0x0000_1FFF

Fig. 7.8 Address editor

7.3.3 *Address Editor*

The *Address Editor* tab provides the slave address mapping corresponding to the master interface. However, please note that the Address Editor tab only appears if the block design contains an IP block that functions as an AXI master interface (such as the MicroBlaze processor) or if an external bus master (outside of *IP Integrator*) is present.

As can be seen in Fig. 7.8, the data and the instruction cache of the MicroBlaze are respectively mapped to the block RAM and local memory, the address of whom is depicted on the offset address.

- Address Map

  Master interfaces reference an assigned memory range container called *address spaces*. Slave interfaces reference a requested memory range container called a memory map. The address space names are related to the usage by the master interface to which it corresponds to. It represents which slaves are mapped to which address space of the master. The entire address map is available in a separate tab called an *Address Editor* tab within the IPI layout.

7.3.4 Parameter Propagation

While designing with IPs in block design, it is important that the configuration user parameters are propagated to the IP blocks connected. It enables an IP to auto-update its parameterization based on how it is connected in the design. For example, the clock frequency set in one of the IP blocks gets propagated through the design. IP can be packaged with specific propagation rules, and *IP Integrator* will run these rules as the block design is generated. However, if the IP cannot be updated to match properties based on its connection, an error is reported to highlight the potential issues in the design.

7.3.5 Validate Design

Validate design enables you to run a comprehensive design rule check as your design is being consolidated which ensures that the parameter propagation, address assignment as described above, and other aspects of the design creation are correct.

In short it ensures the completeness of the design. You can click on the icon available in either the toolbar pane or in the BD canvas pane to run validation checks.

7.3.6 Generate Design

In order to generate the necessary source files from the various IPs used in the block design which are to be used by synthesis and implementation runs, IPI provides a feature to generate the block design called *Generate Block Design* available in the flow navigator upon successful completion of validation of design. It generates various source files like the HDLs respective to the IPs, constraints, and register level files (for processor if any in BD) associated with the individual IP components. If this option is run before *validate design*, this process will first invoke *validate design*; ensure that there are no DRC in the design and will then generate the respective output products. These outputs can be seen in the Vivado Sources pane. Based on the language setting of the project, the output products will be generated accordingly (provided the IP is packaged accordingly).

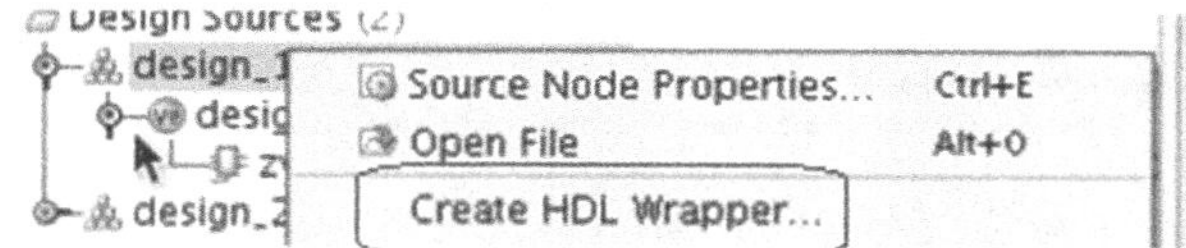

Fig. 7.9 Creating an HDL wrapper

7.3.7 Top-Level RTL Wrapper

The block design can be either the topmost level of the design or it can be instantiated in an RTL which then can be the top level of the design. If the block design is the topmost in the hierarchy of the IPs, IPI provides a way to generate the RTL wrapper for the same which is used in the synthesis flow as shown in Fig. 7.9.

Based on whether the project settings have been set to either Verilog or VHDL, the top-level RTL wrapper will be generated in a respective HDL.

7.3.8 Export Hardware Definition

This feature allows you to transfer the hardware design information to the *Software Development Kit* (*SDK*). It is mainly useful in a hardware-software ecosystem. Usually in a processor-based system, when there is a programmable logic (PL) also present in the design, the hardware definition is exported after bitstream generation which thus includes the complete hardware and software configuration. However, in some cases when there is no PL present, there is no need to generate bitstream, and the hardware information can be exported right after generation of output products.

7.3.9 Creating an Example Design

Vivado provides a way to ensure that you get started with some reference design created in IPI. It has a predefined set of example projects being embedded which can be created at the beginning to the project.

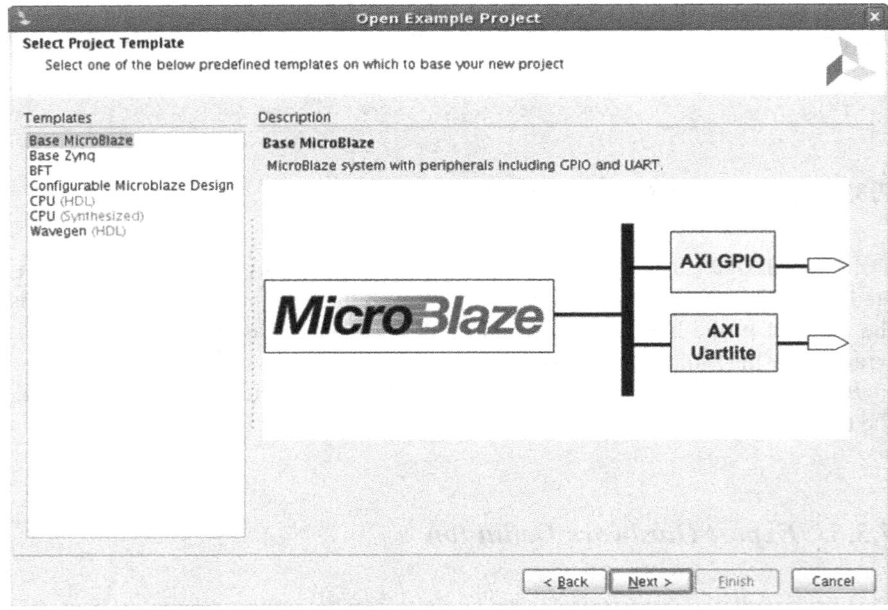

Fig. 7.10 Creating an example design

Figure 7.10 shows availability of reference designs which are available for you. Based on the selection made, the tool generates an IPI-based design which acts as a template design for you. You can alter this design, based on your requirements.

7.4 Tcl Support

One of the powerful aspects of IPI is the extensive backend Tcl support. All the features of IPI can be accessed using a set of Tcl commands which can be executed in batch mode as well as in the GUI mode. Section 2.2 explains more about making use of Tcl support.

Chapter 8
SysGen for DSP

Arvind Sundararajan

8.1 Introduction

DSP systems generally transform a stream of sampled inputs to a stream of sampled outputs. The sampled inputs are usually the result of an analog to digital converter sampling the output of a sensor. Development of a DSP algorithm involves transforming these input samples using numerous mathematical operations like convolution or performing Fast Fourier Transform (FFT). The Implementation of these algorithms requires many visualization aids like plotting the spectra, l density of the output signal or creating scatter plots of complex signals. Development of these systems and algorithms on an FPGA using traditional RTL techniques is very labor intensive due to the lack of libraries to create domain-specific stimulus generators and visualizers. Much of the time would be spent simply creating test benches that try to emulate the deployment environment (Fig. 8.1).

MathWorks tools, in particular *Simulink*, are used for modeling the environments in which DSP algorithms operate. Ability of the Simulink engine to handle models that operate on discrete sample time boundaries as well as continuous signals makes it very easy to graphically model DSP systems and real-world physical signals on the same canvas. Built-in stimulus generators as well as visualizers alleviate the laborious task of creating test benches for DSP system. Synergy between the MATLAB language and Simulink is particularly highlighted in being able to parameterize Simulink blocks and hierarchies using MATLAB expressions.

System Generator for DSP (introduced in 2001) is a high-level modeling and design tool used for implementing designs on Xilinx FPGA. System Generator embeds as a blockset and set of services within MathWorks Simulink. It was the first product of its kind using which DSP System Designers familiar with MATLAB and Simulink could implement their algorithms on Xilinx FPGA. For the first time,

A. Sundararajan (✉)
Xilinx, San Jose, CA, USA
e-mail: asrajan@gmail.com

S. Churiwala (ed.), *Designing with Xilinx® FPGAs*,
DOI 10.1007/978-3-319-42438-5_8

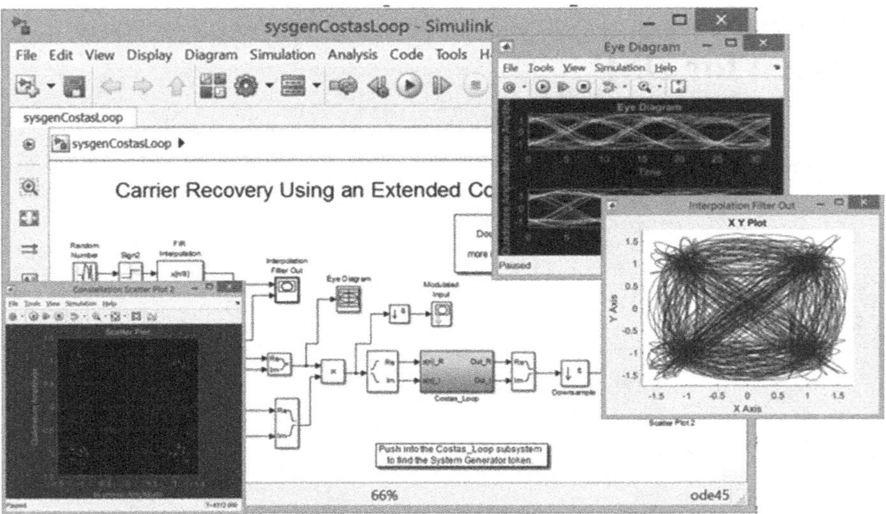

Fig. 8.1 Simulink design environment with visualizer for eye diagram and scatter plot

engineers with no prior FPGA design or RTL experience could get a design running on an FPGA in a matter of hours using System Generator's Hardware Co-simulation.

8.2 Designing in System Generator for DSP

System Generator is different from all other Xilinx tools in that it embeds in a proprietary third-party design environment called Simulink. The access to the tool is provided through a catalog of blocks available in the Simulink library browser as shown in Fig. 8.2. The process of designing is through drag and drop of built-in blocks on the drawing canvas and connecting input and output ports. All System Generator blocks can be distinguished from Simulink blocks due to the presence of the Xilinx logo.

8.2.1 Xilinx System Generator Blockset

The blocks provided for designing and implementing a system (or a portion thereof) are organized into libraries. Basic blocks that model mathematical operations like addition, multiplication, division, etc., are collected together in the Math libraries. Abstractions of Xilinx DSP IP like FIR Compiler and DDS Compiler are available

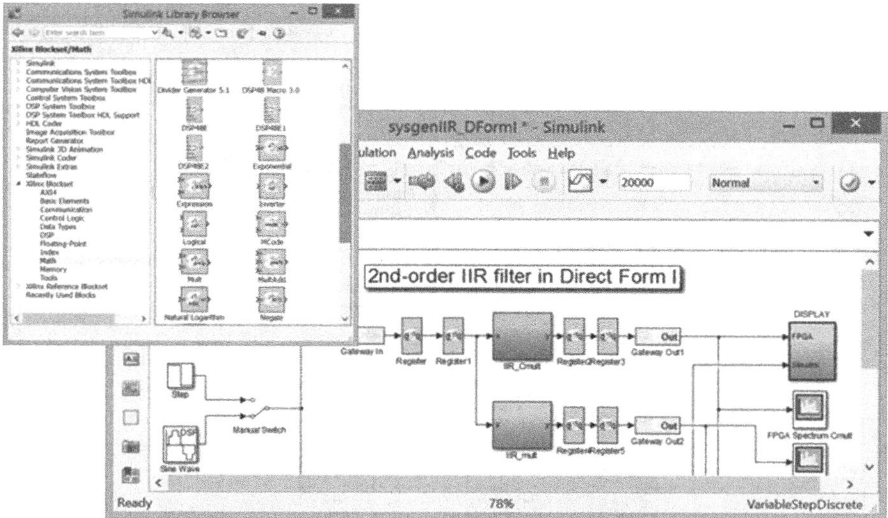

Fig. 8.2 Xilinx blocks with X watermark to distinguish from Simulink blocks

through the DSP Libraries. Blocks that provide rich functionality like Viterbi decoder help accelerate design creation, while low-level blocks like *AddSub* allow users to customize their algorithms. In wireless communication applications, clock speeds are often as high as 491 MHz. Many of these blocks, therefore, expose parameters that can be tuned to achieve these speeds. For example, the Mult Block allows user to tune the latency which helps with pipelining the multiplier. Another option provided on the *Mult* Block is to use embedded multipliers (same as DSP48 slice) which help to close timing at higher clock speeds than implementing it on fabric (Fig. 8.3).

8.2.1.1 Gateway In and Gateway Out

The Xilinx Blockset contains two special blocks called *Gateway In* and *Gateway Out* that mark the boundary between the portion of the Simulink model that forms the test bench and the portion of the Simulink model that forms the design under test. All System Generator blocks must be driven by other System Generator blocks except *Gateway In* which is driven by Simulink blocks. All System Generator blocks can only drive other System Generator blocks except *Gateway Out* which can drive a Simulink block

In Fig. 8.4, *Gateway In* brings data from one of Simulink's stimulus generator (source) block called *Sine Wave*, and *Gateway Out* connects the output of the design to a Simulink visualizer called the *Scope* block.

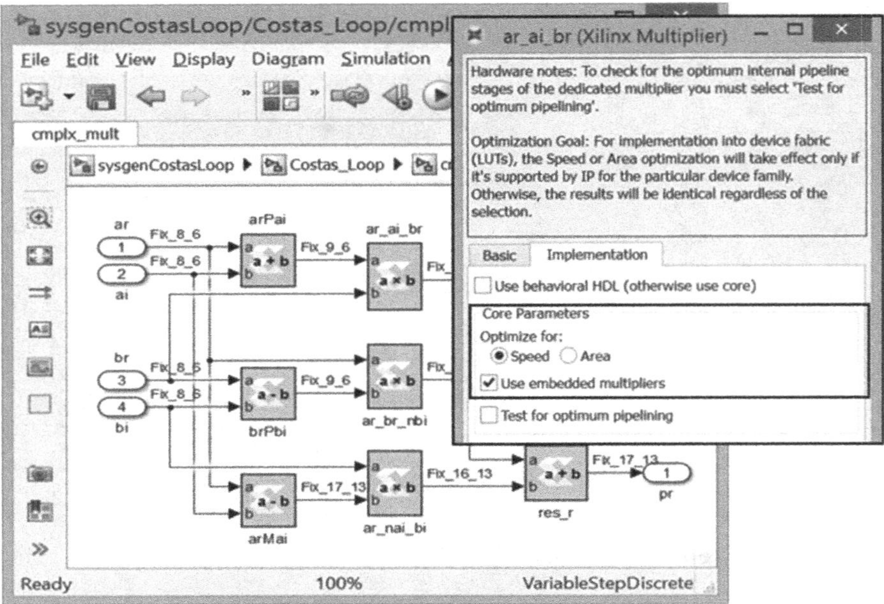

Fig. 8.3 System Generator blocks can be parameterized to extract maximum performance

8.2.1.2 System Generator Token

The System Generator token is a special block in the Xilinx Blockset library that holds information about the System Generator model. It captures project information such as *compilation targets* including *IP Catalog* and *Hardware Co-simulation* (Sect. 8.3.2), the top level HDL (VHDL or Verilog) to be used, Xilinx *Part* to be used, *Target Directory* in which the results of compilation should be placed, etc. The first tab of the user interface for System Generator has been reproduced in Fig. 8.5.

8.2.2 Sample Times and Cycle Accuracy

Simulink provides a generic framework for sample time propagation that can be used to model a variety of different continuous and discrete time systems. All System Generator for DSP blocks except the *Gateway In* only accept and propagate discrete sample times. A discrete sample time is a double-precision number that specifies a time step in seconds. It can either be associated with signals or blocks. The discrete sample time associated with a block tells the *Simulink* engine when to update the outputs and the internal states of the block.

Most System Generator blocks specify the output sample times as function of the input sample times and block parameterization. In general (with some nota-

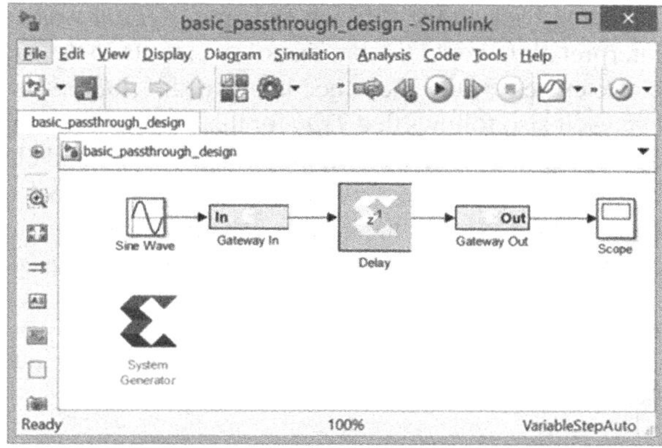

Fig. 8.4 Gateway In and Gateway Out blocks mark the boundary of the design under test

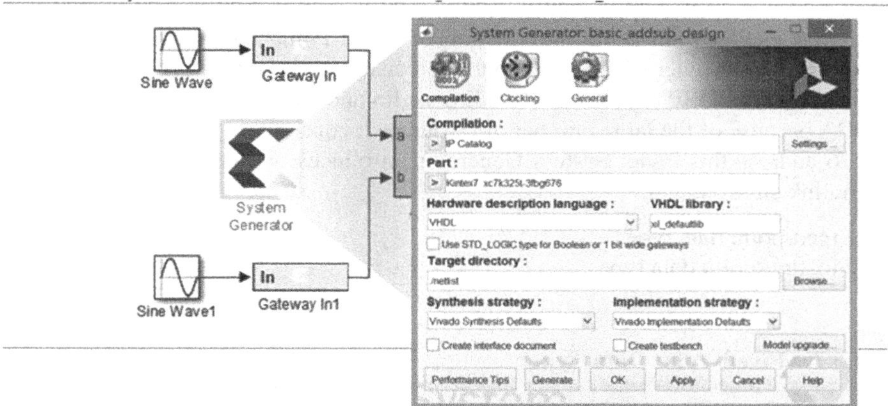

Fig. 8.5 System Generator token and its user interface

ble exceptions), the Simulink engine executes blocks at time steps that is a multiple of the greatest common divisor of the sample times of all the inputs and outputs. For example, if the sample times of the inputs of a Xilinx *AddSub* block are 2 and 4, then the *AddSub* block will specify the output sample time to be 2. This means that at least one of the inputs change at 0, 2, 4, 6, 8 … seconds and output also changes at 0, 2, 4, 6, and 8. Between these times the values on the signals are held constant. This is a very important abstraction that helps with hardware design.

Most digital designs make use of a *clock* that keeps time. A cycle refers to a unit of time representing one clock period. The System Generator token has a field called Simulink *System Period* (*Tsim*) which accepts a double-precision number. This number relates the time in simulation with the time in hardware. An advance in time in simulation equivalent to *Tsim* corresponds to an advance in time in hardware of one clock period.

Going back to the *AddSub* example, if *Tsim* was set to 1, a hardware designer's interpretation would be that one of the inputs to the *AddSub* is held constant for two clock cycles, while the other input to the *AddSub* is held constant for four clock cycles. It also follows that *Tsim* for this particular example cannot be greater than 2 as any signal in an idealized (ignoring logic delays and clock skews) synchronous clock design cannot change between rising edges of clocks. In essence, *Tsim* time steps represent behavior of the system just after each rising edge of the clock in hardware. All Xilinx System Generator blocks provide a cycle accurate simulation behavior.

8.2.3 Data Types

The default built-in data type in Simulink is *double*, and many of the original built-in stimulus generators were only capable of producing double-precision data types (this has changed since the introduction of Fixed-Point Toolbox now known as Fixed-Point Designer). However, double precision is unsuitable for implementing many common DSP algorithms including audio and video signal processing in an FPGA because of the large number of resources it consumes.

To address this issue, System Generator introduces two new data types in the Simulink environment:

- Fixed-point data type
- Floating-point data type

8.2.3.1 Fixed-Point Data Type

Fixed-point data-type format is characterized by the number of bits and binary-point location as shown in Fig. 8.6.

In many digital designs, interpretation of bits as numbers is important; however, it is not necessary to have a large dynamic range that double precision offers. For example, the output of a counter limited to a count of 15 can be represented using

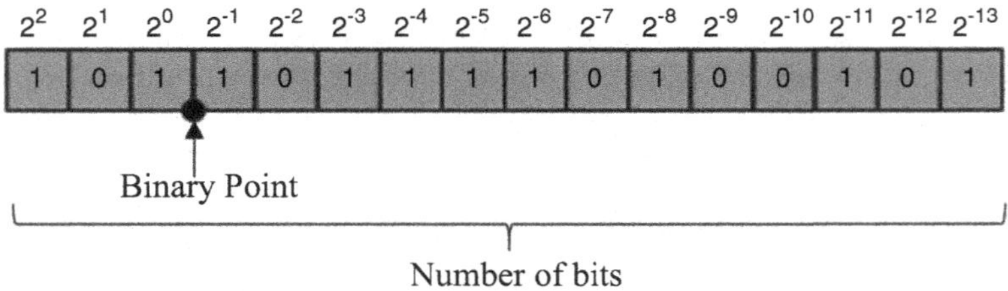

Fig. 8.6 Bit layout of an example fixed-point number with number of bits set to 16 and binary point located at 13

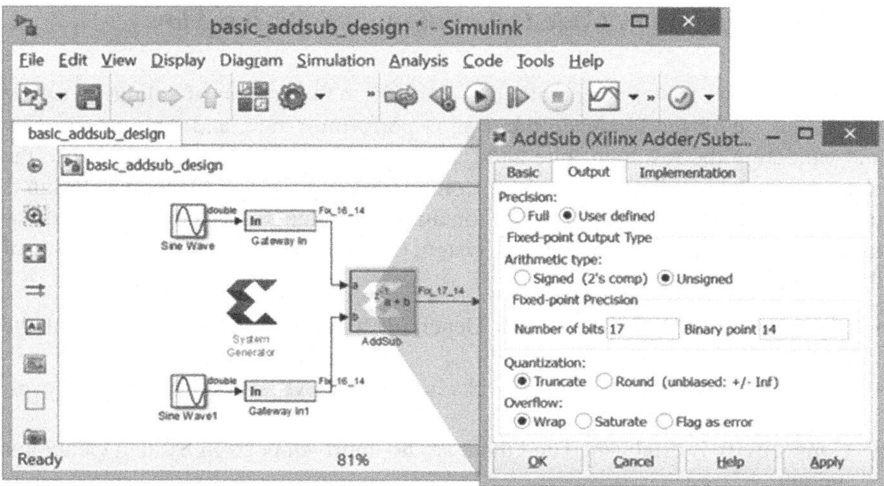

Fig. 8.7 Specifying output type on the AddSub block to keep bit growth under check

four bits. Modeling values as fixed-point data type introduces quantization errors and loss of dynamic range (when compared to double precision), but the resource saving afforded by fixed-point type far outweighs these drawbacks.

Most of the System Generator blocks operate on fixed-point data type and on compilation propagate binary-point alignment as well as bit growth to ensure no loss in precision. For example, if an *AddSub* block configured as an Adder has two 16 bit inputs with binary points located at 8, by default the output bit width would be set to 17 with binary-point location at 8. This can have an undesired increase in resources due to bit growth. Therefore, many of these blocks also have options to specify what the output type should be to keep the bit growth in check. Figure 8.7 shows the user interface to specify the output type on some of the typical operators that show bit growth.

8.2.3.2 Floating-Point Data Type

System Generator also supports floating-point data type including double precision (64 bits), single precision (32 bits), as well as arbitrary precision. This is particularly useful in converting golden Simulink models to System Generator models as well as developing applications that require high dynamic range, for example, matrix inversion. To exploit the flexibility offered by FPGAs, arbitrary precision floating-point data types are also supported wherein the user can explicitly manage the number of bits for exponent as well as mantissa.

Note that the support for floating-point data types is not as extensive as fixed-point data types (i.e., only the blocks in the *Xilinx floating-point library* actually support floating-point data type), and conversion from floating-point block to fixed-point block is supported using the convert block.

8.2.4 Compilation, Code Generation, and Design Flow

Compilation of a System Generator design refers to the process of validating all the parameters on all the blocks in the design, performing rate, and type propagation and validating the results of rate and type propagation. This can be likened to the elaboration phase in HDL compilers. Compilation of a System Generator design is invoked anytime the Simulink model containing a System Generator design is simulated or code generation through the System Generator token is invoked. As part of the compilation process:

- The tool confirms that a System Generator token is placed at the root of the System Generator module.
- User parameters on each of the blocks including MATLAB expressions are evaluated and validated.
- Connectivity is established and there are no undriven ports in System Generator module.
- The type and sample time propagation engine is invoked, and sample times of each block and signals as well as types of the signals are resolved.

Code generation refers to the process of transforming the Simulink model containing a System Generator subsystem into RTL and IP that can be synthesized. Following compilation either simulation can be performed or code generation can be performed. In general, if the compilation is successful, it should be possible to perform code generation or simulation. The general design flow is presented as a flowchart in Fig. 8.8.

8.3 Verification of System Generator Design

8.3.1 RTL Test Bench

Along with the RTL and IP that represents the System Generator design in Simulink, you can also optionally generate a self-verifying RTL test bench. On invoking code generation, the design is simulated in Simulink where *Gateway In* and *Gateway Out* blocks log the data they consume and produce into files with *.dat* extensions. Running the RTL simulation of the test bench uses the data files generated from *Gateway Ins* as stimuli and compares the results of RTL simulation on the output ports of the module with the data files generated from *Gateway Out*. The RTL test bench can be reused to verify the results of synthesis of the System Generator module as well as implementation.

System Generator modules are generally a submodule of a larger design, typically a DSP data path. The RTL test bench allows users to verify the System Generator module in isolation. Also the RTL test bench along with a System

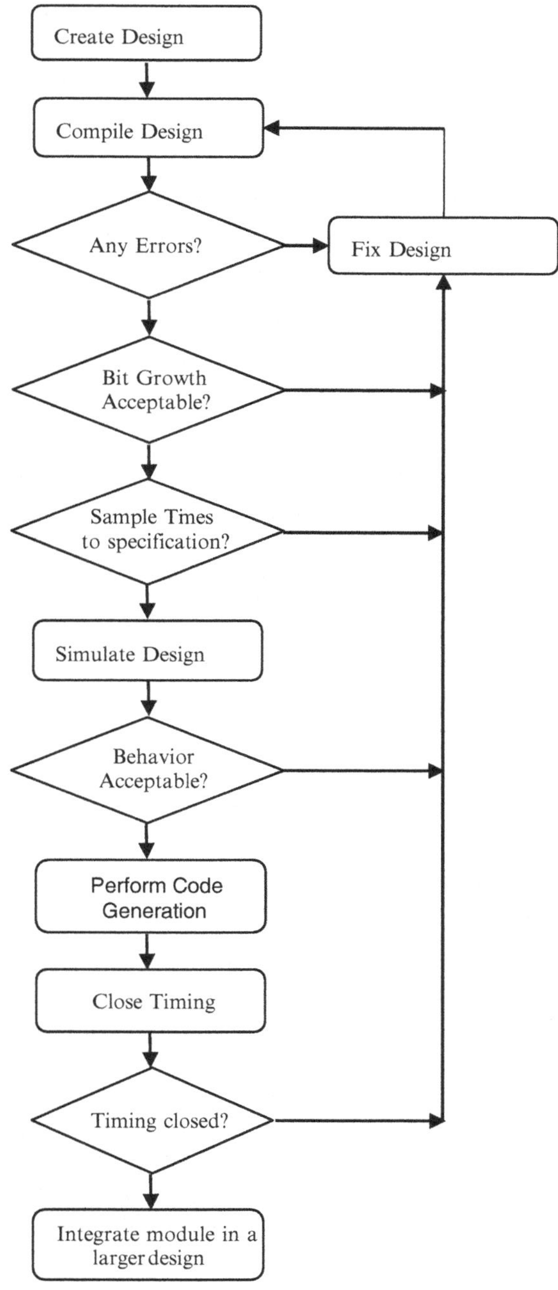

Fig. 8.8 General flow to close on a System Generator module

Generator module provides a handoff mechanism to an RTL developer responsible for integrating the system. This will help the RTL developer to become familiar with the System Generator submodule of a larger design.

8.3.2 Hardware Co-simulation

Many DSP algorithms require processing of a very large number of samples for ratification. For example, a receiver in a communications pipeline may need to process millions of samples to generate BER numbers. Often times even cycle accurate simulation in Simulink may take many days of simulation to verify that the algorithm is correct. To help with reducing the simulation run times, System Generator also has an important feature called *Hardware Co-simulation* that accelerates Simulation by using one of Xilinx's Hardware Evaluation Boards.

To use Hardware *Co-simulation* the design must be compiled for a specific target board. This is done by setting the compilation target for Hardware Co-simulation.

Invoking code generation compiles the design into a bitstream that includes the user design as expressed in Simulink and a communication interface to pass data from the host PC to the FPGA board. Two types of communication interfaces are supported including JTAG and Ethernet. In general Hardware Co-simulation helps only if the Simulink simulation time is on the order of 6 h or more. This is because for each iteration, the design must first be implemented.

8.4 Integrating System Generator Submodule in a System

System Generator provides facilities and services that enable expression and verification of DSP data paths rapidly. However, a system implemented on FPGA includes more than the DSP data path such as interfacing with memory, bringing data in from sensors through ADCs and IOs or HDMI interface. Other Xilinx tools such as Vivado IP Integrator (Chap. 7) or RTL flow with Pin Planner are more suitable for this purpose. To aid with these user flows, System Generator provides *IP Catalog* as a compilation target as shown in Fig. 8.9.

In this compilation mode, in addition to generating products from System Generator that are synthesizable using Vivado, the output products are also packaged into an IP located in the *IP* folder under target directory. This IP can be used in an IPI project or instantiated in an RTL project. More on IPI is covered in Chap. 7, and on using an IP in an RTL project is covered in Chap. 3.

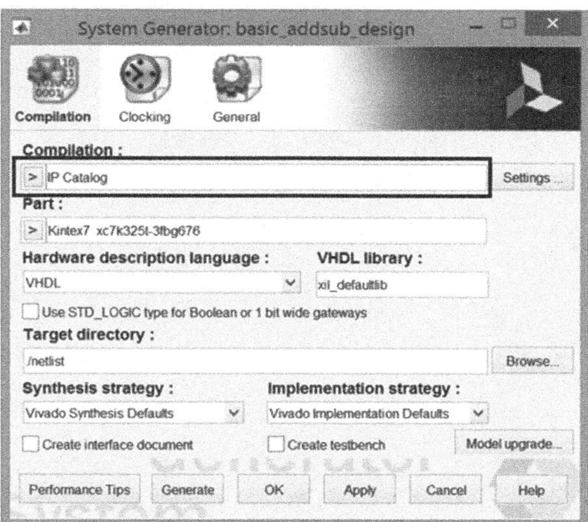

Fig. 8.9 IP Catalog as a compilation target

Chapter 9
Synthesis

Nithin Kumar Guggilla and Chaithanya Dudha

9.1 Introduction

Synthesis is the first step, which maps architecture-independent RTL code into technology-specific primitives. Usually, synthesis tools are supposed to isolate the users from knowing the device details. However, having a good idea of device primitives allows you to fine-tune the synthesis behavior. This might be required mainly for the following reasons:

- Code written for another device might need tweaks in order to get optimal area, performance, and power on the current device.
- Sometimes, synthesis is done on individual parts of the design. So, what might appear as a good optimization decision in the context of that small design might not necessarily be the right decision in the context of the whole design. You might need to guide the synthesis tool in such cases to alter the optimization decisions.
- Sometimes, for designs with special purpose application, you might want to obtain the last bit of performance or area or power—depending on the need—even at the cost of a few other factors.

Synthesis behavior can also have an impact on how efficiently a design can be taken through the back end place and route tools. In the context of this chapter, any synthesis behavior refers specifically to Vivado synthesis tool, though some other synthesis tools may also provide similar capabilities.

N.K. Guggilla (✉)
Xilinx, Hyderabad, Telangana, India
e-mail: gnithin@gmail.com

C. Dudha
Xilinx, San Jose, CA, USA

S. Churiwala (ed.), *Designing with Xilinx® FPGAs*,
DOI 10.1007/978-3-319-42438-5_9

9.2 Designs Migrating from ASIC

9.2.1 Inline Initialization

Each register or latch in any FPGA device can be individually configured to power up in a known *0* or *1* state. This initial state is independent of any asynchronous clear or preset signals used to operate the register. The INIT attribute defines the initial power-up state of registers and latches. An initialization in HDL will cause Vivado synthesis tool to attach the corresponding INIT. The initialization given below will result in register with INIT value as *1*:

```
reg state_bit = 1;
```

In ASIC world, there is no such provision. This would imply that while you need to specifically have a reset or set in an ASIC, for FPGAs, you should just initialize your registers and can get rid of *set/reset* conditions for the flops and latches.

9.2.2 Memory Initialization

FPGAs have dedicated memory blocks (distributed/block RAM). These support initial values which can be used for applications like ROM on power up. Synthesis tools support inferring these initializations when coded using initial blocks and using *$readmemh/$readmemb* to initialize memories:

```
reg [31:0] mem  [1023:0] ;
initial begin
  $readmemb("init.dat", mem) ;
end
```

9.2.3 MUX Pushing

For an ASIC, there is not much difference (in timing) between an adder followed by *MUX* and *MUX* followed by an adder which performs the same functionality. But FPGA architectures have *CARRY* chains which are preceded by LUTs. In this context, consider two scenarios:

- An adder followed by a MUX
- MUX followed by an adder

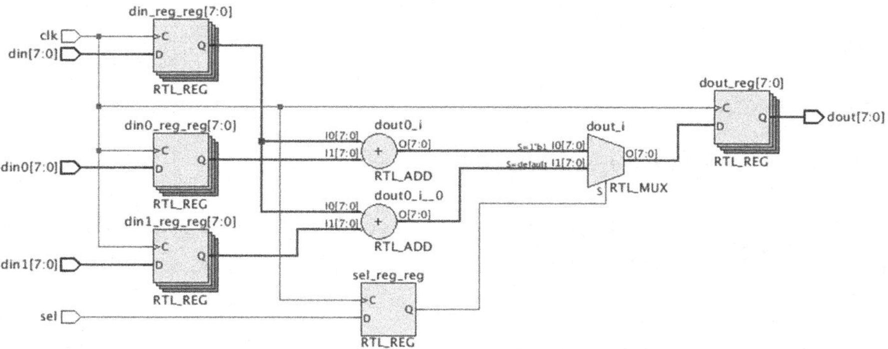

Fig. 9.1 RTL view before MUX pushing

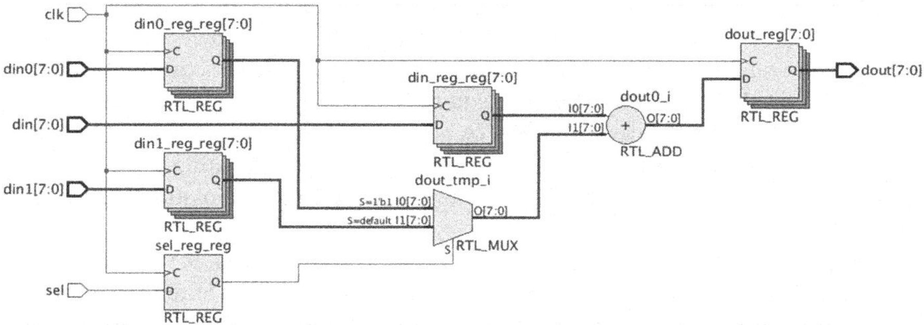

Fig. 9.2 RTL view after MUX pushing

The first one results in 1 additional LUT+1 logic level. The second one can combine the adder into the same LUT. The RTL code segment below would result in an adder followed by a MUX, as shown in the schematic of Fig. 9.1:

```
always @ (posedge clk)
begin
 …..
 if(sel_reg)
      dout <= din_reg + din0_reg;
 else
      dout <= din_reg + din1_reg;
end
```

The critical path in Fig. 9.1 is through the adder and the MUX. This is translated to 1 LUT+2 *CARRY4*+1 LUT. The same functionality can be recoded as below to give a circuit as shown in Fig. 9.2, where *MUX* is moved in front of the adder:

```
always @ (posedge clk)
begin
 ...;
 dout <= din_reg + dout_tmp;
end
dout_tmp = sel_reg ? din0_reg: din1_reg;
```

One-bit addition and the MUX can now be combined into the same LUT. So the critical path is now 1 LUT+2 *CARRY4*.

9.2.4 Clock Gating

Clock gating is a popularly used technique in ASICs and FPGAs for eliminating unnecessary switching activity there by reducing dynamic power. Based on the design functionality, designers will add this gating logic to drive sequential elements which will disable clock as and when required.

Since clocking resources on an FPGA are predetermined, gating might have to be done within the specific clocking structures available. A gating logic on clock path for an FPGA could result in skew and thereby timing violations, especially hold violations.

You can move the gating logic onto *clock buffer* as shown in Fig. 9.3, specially if the same gated clock controls a high number of elements. These clock buffers are designed to prevent spurious clocking, due to change in enable while the clock is in *active* phase.

Gating logic on clock path should typically be moved to *enable* path for flip-flops or latches. Vivado synthesis tool will do this automatically (under user option). But if the structures are too complicated, user intervention might be needed in terms of altering HDL code.

FPGAs have dedicated primitives like block RAMs, DSPs which contribute to a good percentage of the total power. These primitives have *clock enables* which can be leveraged similar to flip-flops if there are clock gating structures on these.

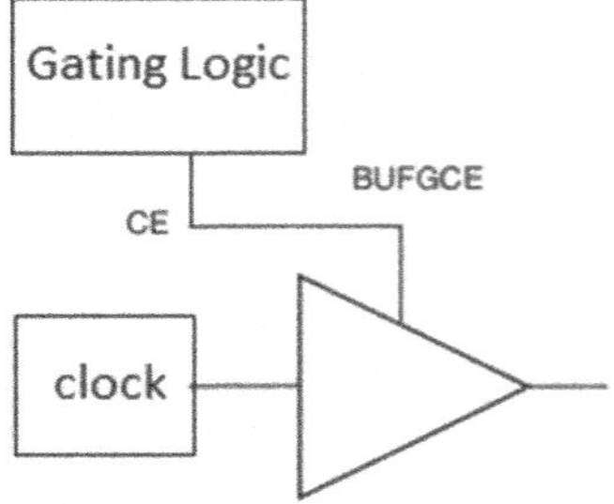

Fig.9.3 Using BUFGCE for clock gating

9.3 Getting the Most of Device Primitives

FPGA is made up of a fixed number of different varieties of structures. Having an understanding of the target architecture and the impact of different mappings will allow you to obtain a very high *QoR*, by tweaking the actual inference and resource mix—depending on your specific design care-abouts.

Same functionality may be realized using different combinations of primitives. If your design makes excessive use of a specific primitive, you might want to implement some of the functionality onto another type of primitive, where possible, even if that other primitive type might usually be considered suboptimal for that specific functionality realization.

This section covers some of the dedicated primitives of Xilinx FPGA and examples of decision-making process to show the best way to obtain optimal results through Vivado synthesis.

The examples given below are w.r.t Xilinx 7 series, UltraScale, and UltraScale+ devices. The basic idea behind providing these is to give a conceptual understanding which can be adjusted for other architectures, depending on the structure available in those future architectures.

9.3.1 SRLs

Xilinx FPGAs contain primitive which is *LUTM* (LUT memory) which can be configured as a sequential element like a shift register (*SRL32*) or a distributed RAM. This section covers some examples to illustrate decision-making process around *SRL*s.

Take a simple example of a delay chain of 64 of 1 bit wide. This can be implemented in 64 flip-flops. These would need at least four slices. Or, they can be implemented in 2 LUTMs—going into a single slice. Each LUT configured as an *SRL32* + an additional flop for better *clock_to_out* which can all go into a single slice.

On the other hand, consider a design having many delay lines with small depth (say 3). If these are mapped to *SRL*, these could cause congestion due to high utilization of *SRLs*. Based on the design statistics, you should control the *SRL* threshold for getting a better trade-off. Vivado synthesis tool provides directives and switches to change the threshold for *SRL* inference.

Structures around SRL also play a role. Consider the following sets of structures:

- Combo logic followed/preceded by SRL
- Block RAM/distributed RAM followed/preceded by SRL
- DSP followed/preceded by SRL

For better *clock_to_out*, synthesis will pull out the last stage of *SRL* into a flip-flop. You can control this behavior using synthesis attributes. You might also consider pulling out the first stage of an *SRL* into flop which would provide higher flexibility for placement. This can be controlled using synthesis attribute *srl_style*. For example, *srl_style* = *reg_srl_reg* will force the tool to have *SRLs with* registers on both sides.

9.3.2 Memories

Designs typically use memories for storing data, buffering, etc. At a fundamental level, a memory is *a bank of flops with decoding logic at the input and MUX logic at the output.*

FPGAs provide dedicated primitives for implementing memories. These are of two types. First is distributed memory which is implemented using *LUTMs* and the second are block RAMs which are hard blocks of size 18 k/36 k.

For very smaller memories, the obvious choice is register based, since the number of flops/glue logic will be less.

For choosing between distributed and block RAM based, the first requirement is synchronous nature. An asynchronous read from the memory will be inferred as a distributed RAM. A synchronous read which implies either output data is registered or the read address being registered is a requirement for block RAM to be inferred.

Since distributed RAM is implemented using a LUT, a six-input LUT can be configured to implement a 64×1 single port memory. A block RAM can support 18 k/36 k bits. Choosing a crossover point on where to use a distributed RAM and block RAM is important. Synthesis tools use thresholds/timing constraints for inferring these memories automatically.

For highly utilized designs where the design is dominant in one of the primitives, i.e., distributed RAMs vs. block RAMs, you should guide the tool using attributes/switches to have a different implementation to get balanced utilization of resources. This will in turn affect the place and route tools on providing better opportunities for placement. There is no deterministic optimal ratio of distributed RAMs vs. block RAMs. The right mix depends on various factors.

Based on few case studies that we have encountered, we will try to mention some of the good practices that can be used based on the scenario. Your design may need its own decision.

9.3.2.1 Distributed RAM Usage

For a highly utilized design with tighter timing constraints, make sure that the distributed RAM percentage of the overall slice usage is relatively low. The reason is that if there are too many distributed RAMs, there would be lot of fabric routing that would converge at each slice/*CLB* which would result in congestion.

Look at configurations of smaller depth, wider data bus. Synthesis tools might look at a combined view of the aspect ratio to decide on inferring distributed or block RAMs. In cases where depth is small, distributed RAMs are a better choice.

For example, depth × width = 32 × 256. This would result in four block RAMs if used in simple dual port (*SDP*) mode. In terms of distributed RAM, it would be 256 LUTs. In this example it is better to go with 256 LUTs. If we look at block RAM bits that are actually inferred, it is 8192 vs. the total capacity of 147,456 (four block RAMs).

9.3.2.2 Block RAM Pipelining

For higher frequencies, always use the pipeline registers or else the *clock_to_out* of the block RAM would limit the performance that can be achieved. In the following situations, synthesis tool might not pull in the register, even if there are pipelines:

- Feedback path on the register
- Fanout from the first stage of the pipeline

Use additional register outside the block RAM for higher performance if block RAM has multi-fanout. Place and route tools would have higher flexibility in placing this register, based on its fanout load placement.

9.3.3 DSPs

DSP blocks come with a number of features. A few to mention are pre-adder, multiplier, and post-adder/accumulator with pipeline register at each output.

This section uses examples based on DSP48E2 from Xilinx UltraScale devices.

DSP48E2 supports a signed multiplier of size 27 × 18, 48-bit post-adder, an input pre-adder which is connected to the 27-bit multiplier port.

9.3.3.1 Extra DSPs Inferred

Note that a multiplier of size 27 × 18 will be mapped into a single *DSP* block only if the inputs are signed. So the first thing to check is if the inputs are unsigned.

Adder followed by multiplier when used for full width will not be packed into a single DSP block. A 27-bit addition would result in 28-bit result and then this 28 bit should be used for multiplication. So, the operand size has grown beyond 27—the width of the multiplier. You need to consider the multiplier input size and calculate the maximum possible at the input of DSP primitive. For a signed multiplier of 27 × 18, taking the *carry* into consideration, the maximum possible adder at the input is 26 bit. If it is unsigned, it would be less by one more bit.

Consider a situation, where the multiplier output is tapped/padded with *0*s. before driving an adder. Multiplier output to adder is hardwired, so if there is some truncation/padding, it cannot be done within single DSP block.

To summarize on the above section, DSP is a powerful block, and to use the capabilities of DSP blocks to fuller extent, make sure you understand the hardwired connections and the widths of the supported primitives internally.

You can make use of DSP's pipeline registers for achieving high performance. Make sure to use all the pipelines if you have a tighter timing requirement.

9.3.4 MUXFs

These are 2:1 Muxes that multiplex LUTs which can be used for implementing wider functions. For example, two LUT6s are muxed by a *MUXF7* which provides a capability for implementing a seven-input function. Similar analogy can be used for MUXF8 and MUXF9. But note that the MUXF8 would have inputs as MUXF7s.

There is always a trade-off of using MUXFs vs. LUT3, for example, to implement a two-input MUX when used in the context of a complete design. Simply specified in another way, if a MUXF is driving a register, then it would be advantageous to use it because there is a direct route from MUXF to register. If it is driving some combo, the LUT3 can be combined with another function which would result in a reduction of one logic level. Synthesis tools can be directed by switches/attributes to control the behavior.

9.3.5 Carry Chains

For implementing arithmetic operations like adder, subtractor, or comparators, dedicated *carry chains* (or, carry look ahead) have faster routes.

When using carry chains, make sure to exploit the capability of the architecture. Avoid using an *adder* and feeding into a combo and then feeding into other *adder*, as shown in Fig. 9.4. In this case though the adders are implemented using carry chains, because of the combo, the exit from *CARRY* to LUT and entry from LUT to *CARRY* will contribute to a larger percentage of the delay. This can be slightly restructured to have adder, adder, combo or combo, adder, and adder (as shown in Fig. 9.5) to minimize the delay.

The other best practice is to use a register at the output of adder so that they can be packed into the same slice.

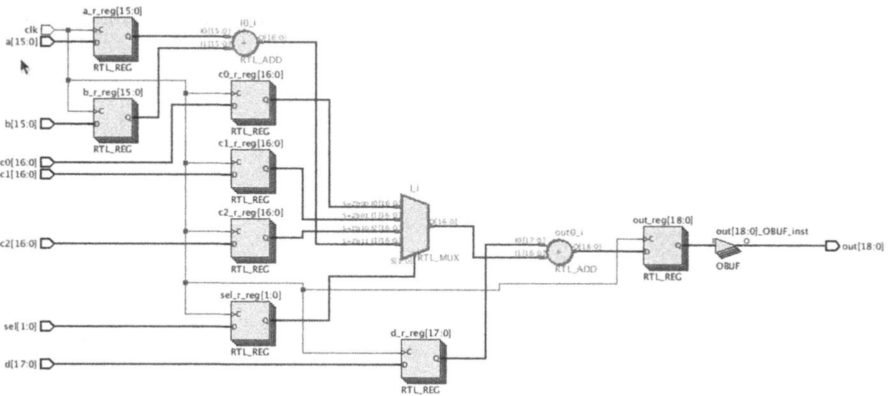

Fig. 9.4 Adder, logic, adder

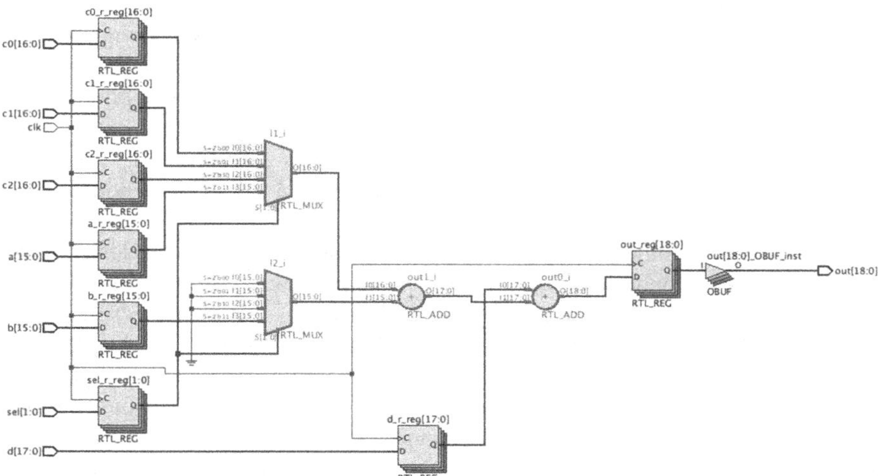

Fig. 9.5 Logic, adder, adder

9.4 Attributes/Directives to Control Synthesis Behavior

Synthesis tools support directives/attributes which can be used in RTL and or XDC to provide finer control to the user. These can be used to change default mapping by synthesis and stop/force some optimizations.

Though a tool could support lot of attributes to control the behavior, an important point to be noted is the implication of these attributes when used in different contexts. Let us look at few examples which illustrate this.

Below is a simple RTL which has *max_fanout* applied on the *enable* signal which drives 1024 flops:

```
module top (
            ....
            output reg [1023:0] dout
           );
(* max_fanout = 10 *) reg en_r;
always @ (posedge clk)
...
  if(en_r)
     dout <= din;
...
```

Consider the scenario when the attribute is not used. This RTL will infer 1025 flip-flops, which would be placed in 65 slices (assuming 16 flops being packed per *slice*). All the 1024 registers have the same control signal.

Now let us consider the case where *max_fanout* of 10 is used. Synthesis will replicate *en_r* 1024/10 (103) times. So we have 103 control sets now. This will use 103 slices for 1024 registers. Due to replication we have 103 additional flops which need ~seven slices.

In the above example, though your intention was to reduce the fanout for improved timing, you can see that control sets played a role which ended up in a considerable area overhead.

Let us look at another example of how multiple attributes when used in conjunction can become nondeterministic. Consider a case where you have an *FSM* and want to force the encoding to one-hot and want to debug this using *logic analyzer*. To achieve this, *fsm_encoding* attribute along with *mark_debug* would be applied.

At a first glance, it looks correct. But there is a conflict. *mark_debug* implies that the exact signal name be intact. With *fsm_encoding* as one-hot, there would be additional flops and state name would get changed. So synthesis tool chooses to honor *mark_debug* and *fsm_encoding* would be ignored. A better way in this case would be to add *mark_debug* post-synthesis via XDC so that the encoded FSM state would be available for debug.

Look at synthesis log file for any message related to attribute being ignored for some reason. *DONT_TOUCH* stops optimizations in the complete flow. So make sure that it is intended.

XDC provides a powerful mechanism which can be used to apply attributes without having the need to change the RTL. Consider a simple example of a module which describes the memory being instantiated in different hierarchies. If you want to map few hierarchies to block RAM and few hierarchies to distributed RAM, a simple Tcl-based XDC can be used, as shown:

```
set_property RAM_STYLE distributed [get_cells u/u1]
set_property RAM_STYLE block [get_cells u/u2]
```

9.5 Synthesis vs. Simulation Mismatch: Common Cases

9.5.1 Global Set/Reset

Vivado netlist simulations do not come out of *reset* till 100 ns. The reason for this is there is a *global set/reset* (*GSR*) in Xilinx FPGAs which retains the initial values on all the flops for the first 100 ns of simulation time.

If you are planning to reuse your testbench from the RTL, ensure that in your testbench, the *reset* is at least asserted for 100 ns before pumping in the actual vectors.

9.5.2 Other Cases

Memories are one area which might expose a difference in RTL vs. synthesized netlist. In cases where the RTL description is mapped to a *simple dual port* or *true dual port* block RAM, during *address collision*, there would be mismatch. Look for warnings during netlist simulation.

In addition, other conventional cases of synthesis vs. simulation mismatches apply.

9.6 Synthesis Switches

Synthesis tools provide switches which act on the complete design. Attributes are for finer control whereas switches are for global control. Let's take a simple example to understand this better. *Flatten_hierarchy* is a switch which has values like *full*, *none*, and *rebuilt*. Let's say you want to flatten the complete design except for few hierarchies. This can be done using the synthesis switch *flatten_hierarchy full* and applying *keep_hierarchy yes* on the desired hierarchies.

These global switches play an important role due to the fact that *place* and *route* tools would see a different view of the same design, depending on the switches used. Though the changes might not be so predominant, factors like *control sets*, *FSM* encoding would result in a change in the resource count and hence different input netlists for *place* and *route* tools.

There are few switches that synthesis tools support to limit the number of inferred primitives like block RAM and DSP. This control is useful, when you are synthesizing a part of the design, and want to leave out resources for other portions of the design. Also for reducing LUT count, the tool can be directed to combine LUTs which will have an impact on timing. So based on the requirement, you can use these switches to fine-tune the output netlist.

9.7 Coding Styles for Improved QOR

RTL coding style plays an important role for getting optimal results. Synthesis tools support specific coding templates for inferring different primitives. Modern tools understand and map to the desired primitives when the user codes in a slightly different way and maintains the intent, but for getting repeatable results, follow the usually recommended coding practices.

Below are few specific suggestions:

- Avoid using too many hierarchies. Else different flattening options will provide significantly different results.
- A simple code using simple constructs is always better. It helps in understanding the intent if you have to revisit the code after a while. Plus, however smart a tool might be, a simpler code would give you the best result always. For example, instead of using a for-loop to assign individual bits, assign the whole bus.
- Look for cases where the tool might do resource sharing. If you need performance, code using parallel structures.
- Avoid instantiation in RTL unless really required. Synthesis tools would not optimize an instantiated primitive.
- Constrain the ranges if the signals/parameters do not need full range evaluation. For example, signals in vhdl if declared as integer type should be constrained as *0–15*, if you need only 16 values.
- Avoid logic functionality while port mapping.

A simple example below illustrates the importance of coding style:

```
module top (
            input [3:0] din,
            output dout
          );
sub u (
       .din(din[3:2] | din[1:0]),
       .dout(dout)
      );
endmodule
module sub (
            input [1:0] din,
            output dout
          );
assign dout = &din;
endmodule
```

In the above example, the output is just a function of four inputs. Synthesizing this one would expect one LUT4 and one logic level. But this may not happen always. Consider a *DONT_TOUCH* on *sub* or this design is run with *flatten_hierarchy none* option. In that case, there would be two LUTs and two logic levels. The point to note here is that coding logic during port mapping can be an easy option for making quick code changes but the repercussions due to the same should be thought up-front.

9.8 Guidelines to Get Best Results Out of Synthesis

- Understand your target architecture, so that you can fully exploit all its capabilities.
- Avoid using too many attributes that could hinder synthesis optimizations. Few of them are like *DONT_TOUCH/MARK_DEBUG*. Debug comes with the cost of additional area/timing penalty. So make sure you understand the intent.
- Look at log file for synthesis info/warning messages, mainly on attributes and if any pipeline registers for block RAM/DSPs are missing.
- There is a misconception that heavy pipelining would make the design meet timing easily. This might have an adverse impact. The reason is too many registers would make packing difficult. Maintain a good ratio of LUT to register, in the range of 1.5. If the ratio is less, relook if pipelining is more than needed.
- Look at logic level distribution post-synthesis. If there are too many paths at the higher side, use a systematic approach to distribute these. Few of the tricks learnt in this chapter should come handy.

Chapter 10
C-Based Design

Duncan Mackay

10.1 Introduction

Recent advances in design tools have enabled a new approach to FPGA design, C-based design. Designing in C allows you to specify your designs at higher levels of abstraction than traditional RTL and obtain the productivity benefits of working at a higher level of abstraction: faster design capture, faster design verification, faster design changes, and easier design reuse.

Figure 10.1 provides an overview of the C-based design flow. The key steps are as follows:

- C simulation verifies that the C function gives the desired behavior.
- High-level synthesis (HLS) is used to synthesize the C function into an RTL design which satisfies the specified performance, timing, and resource requirements.
- RTL verification confirms the output from HLS matches the functionality of the original C function.
- During IP integration, the RTL output from HLS is incorporated into an RTL design.
- RTL synthesis and Place & Route then create the bitstream used to program the FPGA.

The productivity benefits of a C-based design flow are achieved at different stages of the design flow. During the initial development, the primary productivity benefit is provided by fast C simulation which allows you to quickly verify the intended functionality. For example, to simulate a full frame of HD video for a typical video algorithm using C simulation typically takes less than a minute. Simulating the RTL design to perform the same function typically takes a day, if not longer.

D. Mackay (✉)
Xilinx, San Jose, CA, USA
e-mail: duncanm@xilinx.com

S. Churiwala (ed.), *Designing with Xilinx® FPGAs*,
DOI 10.1007/978-3-319-42438-5_10

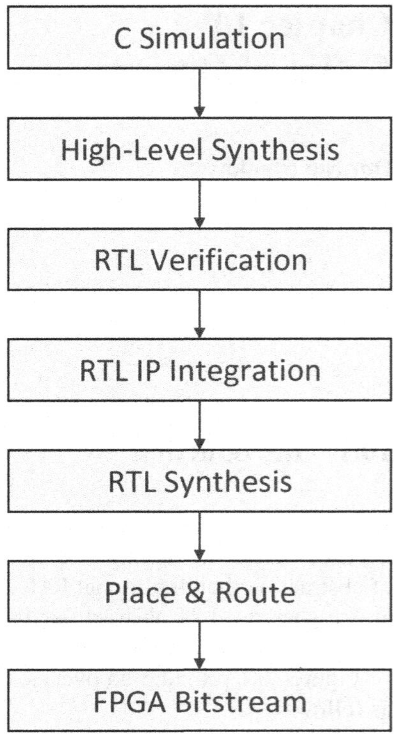

Fig. 10.1 C-based design flow

Once the functionality of the C code has been confirmed, *HLS* allows you to quickly create different RTL implementations from the same C source code, allowing you the time to find the most optimal implementation which satisfies the design requirements: in some cases it may be a fast design at the cost of size, and in other cases it may be a smaller design at the cost of speed (or any point in between).

Once the design is complete, *HLS* allows the same C code to be easily targeted to a different technology or to a different clock frequency or to a different set of performance characteristics, making design migration and evolution substantially easier.

10.2 C Simulation

C simulation is the process of compiling and executing the C program and is the most underappreciated part of a C-based design flow. The benefits of C simulation can be summarized as speed, speed, and speed. It is while performing C simulation that you actually design—create an algorithm, simulate the algorithm, review the results, refine the algorithm, simulate the algorithm, review the results, etc. The fast compilation and execution times of C simulation allow these design iterations to be performed quickly and productively.

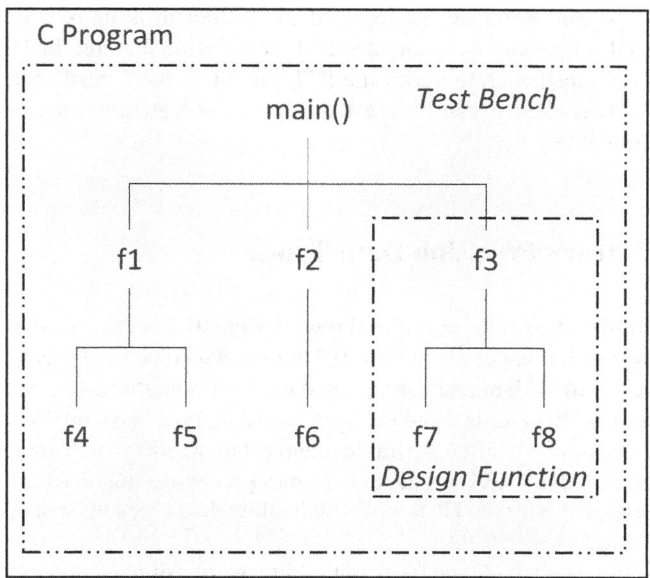

Fig. 10.2 Testbench and design function

As highlighted in Fig. 10.2, the top level of every C program is the *main()* function. In a C-based hardware design flow, the C program is considered to be two separate components, the C testbench and the design function to be synthesized into hardware.

In the example in Fig. 10.2, the C program contains eight sub-functions, *f1–f8*. Function *f3* is the top-level function for synthesis, and everything below function (including) *f3* is the design function to be synthesized (functions *f3*, *f7*, and *f8*). The C testbench is everything below (including) the level *main()* excluding the design functions (functions *main()*, *f1*, *f2*, *f4*, *f5*, and *f6*). The testbench creates input for the design function and accepts output from the design function and hence is used to verify the design function.

A key part of any productive C-based design flow is an intelligent testbench: one which both analyzes and verifies the results from the design function. Figure 10.3 shows an example C design and testbench. The design function, shown on the left-hand side, is a simple design which reads a set of input data from array *DataIn* and determines the minimum and maximum values in the data set. The C testbench, shown on the right-hand side, creates a set of input data, calls the design function, analyzes the output results from the design function (in this simple example, by comparing them to the expected results), and sets the return value to *main()* as zero only if the results are correct.

In a more complex design than the example shown in Fig. 10.3, the input data may be read from a data file on the disk, and the output results may be compared against golden results also read from a data file or from results generated in the testbench. The concept however is the same. C simulation is used to exhaustively verify the design and the C testbench is used to exhaustively analyze the results.

A final important point on the topic of the testbench is its re-use later in the design flow. HLS provides an automated RTL verification feature: the HLS tool will generate an RTL testbench to verify the RTL output. If the C testbench checks the results, an RTL testbench can be created which automatically checks the results after RTL simulation.

10.3 Arbitrary Precision Data Types

An interesting feature of the example shown in Fig. 10.3 is the use of arbitrary precision data types (the header file in Fig. 10.3 defines data type *data_t* as *ap_int<12>*). All data types in the C language are defined on 8-bit boundaries—a *char* is 8-bit, a *short* is 16-bit, an *int* is 32-bit, and the *long long* data type is 64-bit. When performing hardware design, it is often desirable to have data widths which are more precise than those provided in the C language. Arbitrary precision data types are a library of data types provided with the HLS tool which allow data types to be specified in any size, from 1-bit up to 4096-bit.

The most obvious advantage for using arbitrary precision data types is synthesis. If the design needs to read 18-bit data and perform 18-bit multiplications, it is a waste of hardware resources to use the next largest data type in the C language, the *int* data type (32-bit) which results in 32-bit multipliers and 32-bit registers. Arbitrary precision data types allow the hardware designer to accurately size data values in the C function and ensure the most optimal hardware is created.

A less appreciated benefit of arbitrary precision data types is that they allow accurate C simulation to be performed before synthesis. If specific (signed or unsigned) data types are required in the design, they can be both specified and verified in the C function before synthesis. Fast C simulation is used to confirm the bit-accurate behavior.

minmax_frame.cpp

```
#include "minmax_frame.h"
void minmax_frame(data_t  DataIn[N],
                  data_t *min,
                  data_t *max)
{
  data_t t_min, t_max;
  Loop1: for (int i=0;i<N;i++) {
    if (i==0) {
      t_min = DataIn[i];
      t_max = DataIn[i];
    } else {
      t_min = (DataIn[i]<t_min ? DataIn[i]:t_min);
      t_max = (DataIn[i]>t_max ? DataIn[i]:t_max);
    }
  }
  *min = t_min;
  *max = t_max;
}
```

minmax_frame.h

```
#include <stdio.h>
#include "ap_int.h"

#define N 8
typedef ap_int<12> data_t;

void minmax_frame(data_t DataIn[N],
                  data_t *min,
                  data_t *max);
```

minmax_frame_test.cpp

```
#include "minmax_frame.h"
int main () {
// Specify input data set
  data_t Input_Frame[N]={12,22,-3,-65,0,34,93,0};
  data_t max_out, min_out;
// Specify expected results
  data_t max_expect = 93;
  data_t min_expect = -65;
  int retval = 1;

// Call the function to operate on the data
  minmax_frame(Input_Frame,&min_out,&max_out);

// Analyze the results
  if ((max_out == max_expect) && (min_out == min_expect)) {
    printf("Results Match: ");
    retval = 0;
  } else {
    printf("Results Differ: ");
    retval = 1;
  }
  printf("==> min=%d; max=%d;\n", min_out.to_int(), max_out.to_int());

// Report the Results
  if (retval != 0) {
    printf("Sim failed  !!!\n");
  } else {
    printf("Test passed !\n");
  }
  return retval;
}
```

Fig. 10.3 Minmax_frame design example

10.4 High-Level Synthesis

An overview of the HLS process is shown in Fig. 10.4. A key difference between RTL design and C-based design is that HLS synthesizes a single top-level C function into an RTL design which is then incorporated as an IP block into a larger RTL design. The C function is never the top level of an FPGA design; rather HLS is used to quickly create RTL IP blocks which are then assembled in an RTL environment.

The inputs to HLS are the C design function, a C testbench to verify the behavior of the C design function, and design constraints and optimization directives to specify the performance and structure of the RTL design.

The design constraints are the target technology and the clock frequency. The target technology specifies the component delays. Given the component delays and the clock frequency, HLS creates an RTL design which meets timing after RTL synthesis. HLS determines how much logic can be executed in each clock cycle and then creates an FSM to sequence the design operation. It can be expected that when targeting a newer and faster technology, HLS is able to perform more operations within a clock cycle, and hence finish in fewer clock cycles, than when targeting an older and typically slower technology.

Optimization directives may be used to specify the performance and area of the RTL design. During the synthesis process, the HLS tool will perform some default optimizations. These defaults are specified in the documentation provided with the HLS tool. Optimization directives are used to create an RTL design with optimizations which are different from the default synthesis, e.g., to vary the area-performance trade-off point.

The outputs from HLS are an RTL design and reports which detail the performance of the design and an estimate of the maximum delays and the resources required to implement the design. At this point in the design process, only estimates of the timing and area are reported—the exact details cannot be known until RTL synthesis and Place & Route are performed—however, the estimates are generally accurate (±10 %).

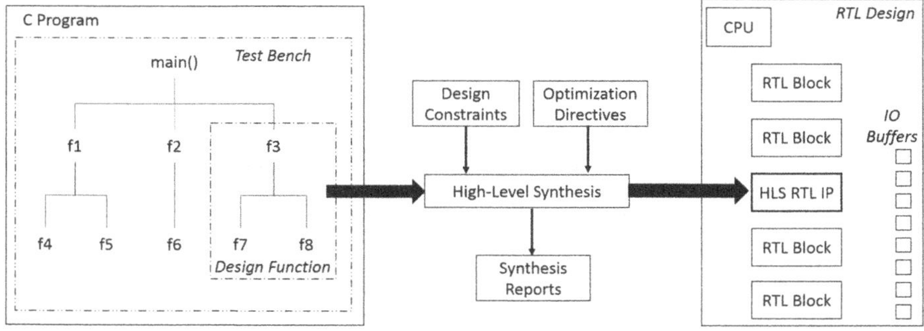

Fig. 10.4 High-level synthesis design flow

10.5 Interface Synthesis

C synthesis may be thought of as two separate processes: interface synthesis and design synthesis (although both are very much intertwined). Interface synthesis is the process of converting the arguments of the design function from simple data values to an RTL cycle accurate interface which optionally may include an IO protocol.

The *minmax_frame* example shown in Fig. 10.3 helps demonstrate the concept of interface synthesis. Figure 10.5 shows the top-level function for synthesis with arguments *Dataln*, *min*, and *max*. After synthesis, these C arguments may be transformed into RTL interfaces shown (in both Verilog and VHDL) in Fig. 10.5.

A clock and reset are added to the RTL design. The tool provides options to control whether the reset port is active-high, active-low, or if it is present at all.

10.5.1 Port-Level IO Interfaces

Each of the data arguments from the C function—*Dataln*, *min*, and *max*—are transformed into RTL data ports with associated interface protocol signals. In this particular case, the array *Dataln* is transformed into a block RAM interface. This interface protocol assumes array *Dataln* is a block RAM outside the design and is therefore accessed with standard block RAM address, data, and chip-enable signals. In this case, HLS determined this port is only ever read and hence there is no requirement for a write-enable (*WE*) port.

Similarly, HLS automatically determined arguments *min* and *max* are only written to and hence these are implemented as output ports in the RTL design. In the example shown in Fig. 10.5, both ports are implemented with an associated output valid signal to indicate when the data is valid.

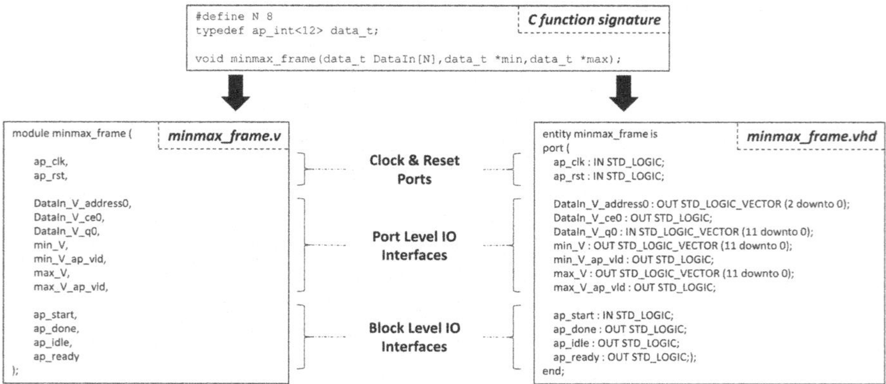

Fig. 10.5 Interface synthesis

Interface synthesis provides many options for interfaces. The array *DataIn* could also be implemented as an AXI master or AXI-Lite interface. If the array is accessed in a streaming manner, with each address location accessed in sequential order (as in this case), the *DataIn* port may be implemented as an AXI-Stream interface, or a FIFO interface, or a two-way handshake interface.

In a C-based design flow, it is highly advisable to synthesize the interfaces with *IO* protocols. This allows the final RTL design to be simply connected to other RTL blocks during the RTL integration phase without you manually trying to determine when the data may be read or written.

10.5.2 Block-Level IO Interfaces

In addition to the port-level *IO* protocols, HLS may optionally add a block-level *IO* protocol as shown in Fig. 10.5. A block-level *IO* protocol is a protocol which is associated with the design or block, rather than any particular port. In Fig. 10.5, the *ap_start* port controls when the block can start its operation, the *ap_ready* indicates when the design is ready to accept new input data, and the *ap_done* and *ap_idle* signals indicate when the design has completed its operation and is idle. Block-level *IO* signals may also be implemented as an *AXI*-Lite interface allowing the RTL IP to be easily controlled from a CPU or microcontroller.

The block-level *IO* and port-level *IO* protocols also help enable automatic verification of the RTL. Given a handshake protocol on both the design and the IO ports, it is always possible to automatically generate an RTL testbench. Without these handshake protocols, it may only be possible to automatically generate a testbench for certain cases. Even if the *IO* protocols are not required for the design, it is worth considering that the small overhead in logic means you do not have to write an RTL testbench.

As with synthesis in general, the HLS tool will have a default interface protocols for each type of C argument (arrays, input pointers, output pointers, etc.). You can then use directives to specify interface protocols other than the defaults.

10.5.3 Interface Options

As noted earlier, HLS provides many options for selecting interface protocols. Figure 10.6 shows some examples of the type of interface which may be created for the *minmax_frame* C code example shown in Fig. 10.3:

A. This is the case shown in Fig. 10.5. The array is implemented as a block RAM interface, and the output ports are implemented with output valid signals.
B. In this case, the array is partitioned into discrete elements and each is implemented as an AXI-Steam interface. Since *N* is 8 in the *minmax_frame* example,

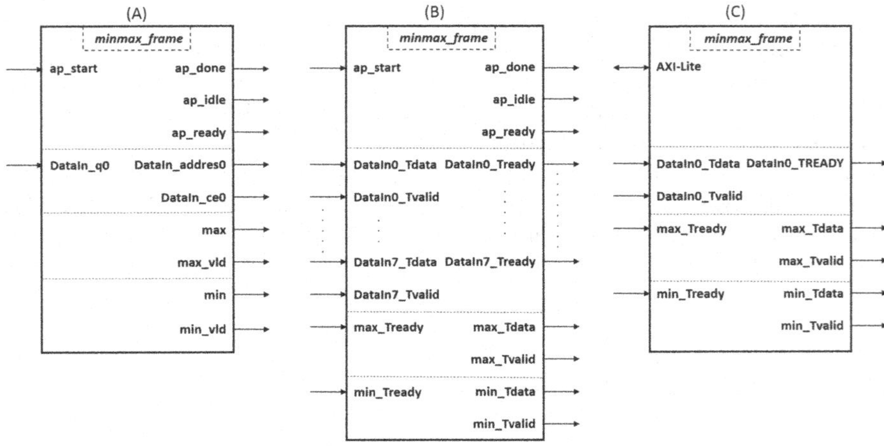

Fig. 10.6 Interface synthesis variations

there are eight discrete ports for the inputs, allowing all inputs to be read simultaneously.

C. In the final case, since all data accesses are sequential in this example, both the input array and the output ports are implemented as AXI-Stream interfaces. Also, in this example the block-level *IO* protocol is implemented as an AXI-Lite interface.

Once you have selected the IO protocols, HLS design synthesis then optimizes the internal logic to maximize the performance of the design.

10.6 Measuring Performance

Performance in HLS is measured by the design latency and *initiation interval* (*II*). Figure 10.7 shows an example design which takes five clock cycles to complete. It starts in state S1 where it performs a read on the data port and proceeds through to state S5 where the output is written.

- The latency is defined as the number of cycles it takes to complete all outputs. In Fig. 10.7, the latency is five clock cycles.
- The *initiation interval* (*II*) is defined as the number of cycles before the design can start to process a new set of inputs. In Fig. 10.7, the next read is not performed until the design has completed, and hence the *II* is six clock cycles.

Both latency and *II* may be specified using optimization directives. Typically, the key performance metric is the *II*: how quickly the design processes new input data and produces output data. In most applications, the goal is to create a design which can read new inputs every clock cycle (*II* = 1).

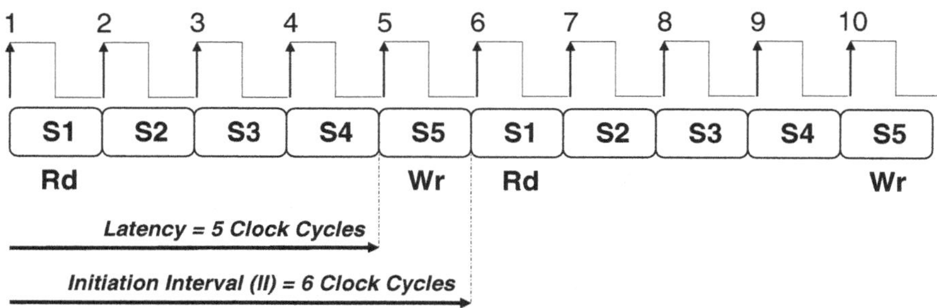

Fig. 10.7 HLS performance metrics

The resources used to implement the design may also be considered a performance metric. HLS provides reports which specify how many LUTs, flip-flops, DSP48, and block RAMs are used. Optimization directives may be used to control the number of these resources; however, doing so impacts the latency and/or the *II*.

10.7 Optimizing Your RTL

When your optimization goals are different from those provided by the default optimizations performed by HLS, you can specify optimization directives to control the RTL implementation. HLS provides a number of optimizations, and it is difficult to review all of the optimizations here; however, it is worth reviewing a few of the key optimizations to provide a sense of what is possible. The *minmax_frame* example shown in Fig. 10.8 can be used to highlight the key HLS optimizations.

10.7.1 Increasing Data Accesses

Arrays are a collection of elements accessed through an index and are synthesized into a block RAM, which is a collection of elements accessed through an address. If an array is on the top-level interface, it is assumed to be outside the design and a block RAM interface is created. Conversely, if the array is inside the C function, it is implemented as a block RAM inside the design.

Arrays may be partitioned and mapped. Partitioning an array splits it into multiple smaller block RAMs (or block RAM interfaces). Since a block RAM only has a maximum of two ports, arrays are typically partitioned to improve data access, allowing more data samples to be accessed in a single clock cycle. Mapping arrays together implements multiple arrays in the C code into the same block RAM, saving resource but often reducing data accesses and limiting data throughput.

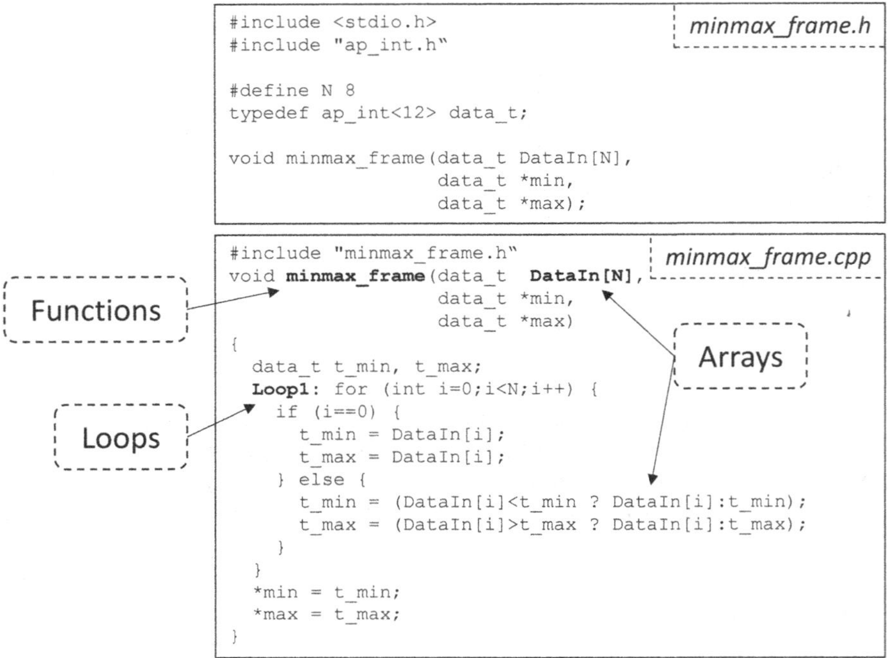

minmax_frame.h
#include <stdio.h>
#include "ap_int.h"
#define N 8
typedef ap_int<12> data_t;
void minmax_frame(data_t DataIn[N],
data_t *min,
data_t *max);
minmax_frame.cpp
#include "minmax_frame.h"
void minmax_frame(data_t DataIn[N],
data_t *min,
data_t *max)
{
data_t t_min, t_max;
Loop1: for (int i=0;i<N;i++) {
if (i==0) {
t_min = DataIn[i];
t_max = DataIn[i];
} else {
t_min = (DataIn[i]<t_min ? DataIn[i]:t_min);
t_max = (DataIn[i]>t_max ? DataIn[i]:t_max);
}
}
*min = t_min;
*max = t_max;
}
Functions
Loops
Arrays

Fig. 10.8 Key optimization objects

In both cases, the array optimizations allow the C code to remain unchanged. Optimization directives are used to instruct HLS to implement the most ideal RTL structure without any need to change the source code.

Loops may be left rolled or they may be unrolled. In a rolled loop, HLS synthesizes one copy of the loop body and then executes it multiple times. Using the *minmax_frame* example from Fig. 10.8, the logic to perform the reads and comparisons is created and then an FSM will ensure the logic is executed eight times (since $N=8$ in this example). This ensures the minimum amount of logic is used, but it can take many clock cycles to complete all operations specified by the loop.

Loops may be partially or fully unrolled. Using the *minmax_frame* example from Fig. 10.8, if the loop is partially unrolled by a factor of, say, 2, this would create two copies of the logic in the loop body and the design will execute this logic ($8/2=4$) four times. This creates more logic than a rolled loop, but now allows more reads and writes to be performed in parallel, increasing throughput (or in other words, reducing the *II*).

At this point, you can perhaps start to see the interaction between the options for interface synthesis and design synthesis:

- Completely unrolling the loop in the *minmax_frame* example creates eight copies of the hardware and allows all reads and writes to occur as soon as possible: potentially, all in the same clock cycle if the frequency is slow enough (or the target technology is fast enough).

- However, if the *DataIn* interface is implemented as a block RAM interface, only a maximum of two reads can be performed in each clock cycle. Most of the hardware is wasted since it must sit and wait for the data to become available at the input port.
- To take advantage of all the hardware created by a fully unrolled loop, the solution here is to also partition the *DataIn* input port into eight separate ports (or four separate dual-port block RAM interfaces).

Similarly, only partitioning the input port does not guarantee greater throughput: the loop also has to be unrolled to create enough hardware to consume the data.

10.7.2 Controlling Resources

Functions and loops represent scopes within a C design function and may have optimization directive applied to the objects within them. A scope in C is any region enclosed by the braces *{* and *}*. Optimization directives may be applied to functions and loops to control the resources used to implement the functionality. For example, if the C code contains 12 multiplications, HLS will by default create as many hardware multipliers as necessary to achieve the required performance. In most cases, this will typically be 12 multipliers.

Optimization directives may be used to limit the actual number of multipliers in the RTL design. For example, if an optimization directive is used to limit the number of multipliers to 1, this will force HLS to allocate only one multiplier in the RTL design and hence share the same hardware multiplier for all 12 multiplications in the C code. This will result in a smaller design, but sharing the resource (the multiplier will have a 12:1 mux in front of it) will mean the design requires more clock cycles to complete as only one multiplication may be performed in each clock cycle.

10.7.3 Pipelining for Performance

Functions and loops may also be pipelined to improve the design performance. Figure 10.9 shows another example of the performance metrics discussed earlier (in Fig. 10.7). In this example the design is pipelined. States *S1* through *S5* represent the number of clock cycles required to implement one execution of a function or one iteration of a loop. The design completes the read operation in state *S1* and starts the operations in state *S2*. While the operations in state *S2* are being performed, the next iteration of the function or loop can be started, and the operations for the next *S1* state can be performed while the operations in the current *S2* state are performed.

As Fig. 10.9 demonstrates, when pipelining is used, there is no change to the latency, which is still 5 as in the previous example (Fig 10.7). However, the *II* is now 1: the design now processes a new data input every clock cycle, for a 5× increase in throughput. Thus, pipelining resulted in this improved performance

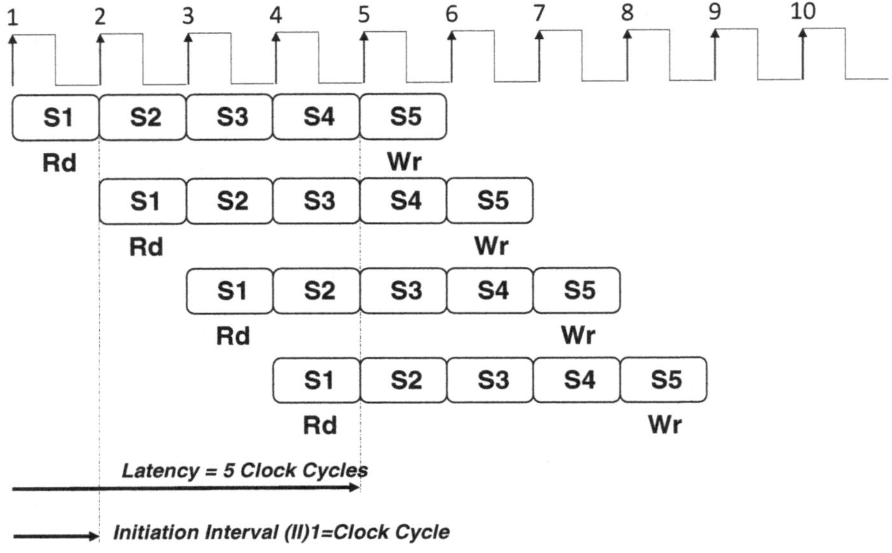

Fig. 10.9 Performance improvement with pipelining

with only a minimal increase in resources (typically a few extra LUTs and flip-flops in the FSM).

Pipelining is one of the most used and most important optimizations performed by HLS.

10.8 Optimization Methodology

Any methodology for creating an optimal RTL implementation ideally requires understanding what the requirements of the RTL implementation are. However, the following methodology assumes you wish to create the highest-performing design. If this is not the case, skip steps 4 and 5:

1. Simulate the C design and ensure the results are checked in the testbench.
2. Synthesize the C code to create a baseline design. This will be the default synthesis performed by HLS and provide you with a starting point for optimization.
3. Apply the optimizations for interface synthesis. This ensures the interfaces are of the required type to integrate the design with the rest of the system.
4. Apply pipeline directives.
5. Address any structural issues which create bottlenecks and prevent pipelining achieving the desired *II*, such as partitioning arrays and unrolling loops.
6. Use the optimization directives which control the allocation of resources to improve the area if this is required.
7. Finally, if the latency is a performance requirement, specify any latency directives.

	Example 1	Example 2	Example 3
Initiation Interval	27	7	8
Latency	26	6	9
LUTs	74	253	123
Flip-Flops	62	225	72
Lines of Verilog	339	724	483
Lines of VHDL	382	806	573

Fig. 10.10 Example design implementations

This methodology can be applied to the *minmax_frame* function to create three data points for design comparison. The clock frequency is specified as 4 ns and a Kintex7 device is targeted.

The performance and resources for the three examples are shown in Fig. 10.10. Remember that this design example has eight input values and therefore the reported *II* is the number of cycles before another eight new inputs can be processed.

Example 1: Small Design

- Input array *DataIn* is specified as a block RAM interface.
- Both outputs are specified with an output valid signal.
- A block-level IO protocol is specified.
- The loop is left rolled.

Leaving the loop rolled ensures the minimum amount of hardware; however, the latency and *II* are the highest because HLS creates logic to implement the body of *Loop1* and then executes the same logic eight times sequentially (calculating each iteration of the loop before starting to calculate the next iteration).

Example 2: Fastest Design

- Input array *DataIn* is completely partitioned into eight separate ports.
- Since the input array *DataIn* is read in sequential order, it is specified as an AXI-Stream to reduce resources (no address generation logic) resulting in eight separate AXI-Stream interfaces.
- Both outputs are specified as AXI-Stream interfaces.
- A block-level IO protocol is specified.
- The loop is fully unrolled.

Unrolling the loop creates a design with the largest amount of hardware—eight copies of the logic required to implement the loop body—and partitioning the input ports allows parallel reads and writes. This creates the fastest design but also uses the greatest number of resource. If the clock frequency is reduced, this design can complete in a single clock cycle.

Example 3: Pipelined Design

- Since the input array *DataIn* is read in sequential order, it is specified as an AXI-Stream to reduce resources.

- Both outputs are also specified as AXI-Stream interfaces.
- A block-level IO protocol is specified as an AXI-Lite interface.
- *Loop1* is pipelined.

Pipelining the loop keeps the hardware to a minimum while still ensuring the design is able to process one sample per clock cycle (*II* = 8: the design can process eight inputs in eight clock cycles).

The ability to generate multiple RTL implementations from the same C code is a large productivity benefit of using HLS. You are able to explore the design space to create the most optimum design implementation.

10.9 A Productivity Data Point

For the *minmax_frame* example used throughput this chapter:

- Writing the C code, the C testbench, and performing C simulation to verify the results took approximately 45 min.
- The run time to generate each of the three HLS solutions shown in Fig. 10.10 is approximately 3 min per solution.
- Between each solution, the time to determine, select, and apply the optimization directives is approximately 5 min.
- Within approximately 1 h, these solutions represent three unique RTL implementations and 1500 lines of RTL HDL code.

Since the clock frequency and target technology are input parameters to HLS, this design may be targeted to a new target technology or a new clock frequency and new RTL generated in a matter of minutes. Although this is a small demonstrative example, the productivity benefits scale when working on larger designs.

10.10 RTL Verification

Automatic RTL verification is a feature of HLS. Since the HLS tool knows the interfaces which are created in the RTL, it is possible to automatically create an RTL testbench to verify the RTL output from HLS. This allows the RTL to be verified without the requirement to create an RTL testbench.

Since the RTL verification is based on the C testbench, the amount of verification which is performed on the RTL is exactly correlated with the effort spent writing a C testbench which exhaustively verifies all modes. As stated earlier, spending time and effort to create a C testbench which exhaustively tests all modes is a productive investment, since C simulation is fast and productive and the investment in this is automatically leveraged into the RTL verification.

RTL verification typically takes substantially longer to complete than any other part of a C-based design flow. It is therefore recommended to only perform RTL verification when the design exploration process is complete or whenever you wish to take a representative sample through the remainder of the design flow.

Verification confirms the behavior of the RTL matches the behavior of the C code simulation. To verify the RTL in the context of the other RTL blocks in the full FPGA design, the RTL output must be integrated into the FPGA RTL design project.

10.11 RTL Integration

The output from HLS is used as RTL input to the remainder of the FPGA design flow. The HLS output is provided in industry standard RTL format (Verilog and VHDL) and in gate-level format (EDIF). The most productive methodology for using the outputs of HLS is the one which uses an IP integration flow where the RTL output from HLS is another IP block in the RTL system along with existing RTL IP.

An IP integration environment allows the IP blocks, including the HLS-generated RTL design, to be easily integrated into the chip-level design, and is explained in Chap. 7. It would typically take more effort to add the HLS IP into the chip-level RTL design manually (connecting each port in a text editor), than using IPI. Since IPI uses IPs based on AXI protocol, you are highly encouraged to use AXI interfaces for your HLS designs, allowing the HLS IP to easily be integrated into your FPGA RTL design using the IP integration environment.

10.12 Tcl Support

The final part of any productive C-based design flow is the use of a Tcl script to take advantage of batch processing. Batch processing through Tcl is supported by HLS, allowing C simulation, C synthesis, RTL verification, and RTL IP integration to be performed efficiently in batch mode.

Chapter 11
Simulation

Saikat Bandopadhyay

11.1 Introduction

Simulation is a way to verify the functionality of design by creating an HDL model and putting it through various input conditions and verifying the output. If the FPGA design doesn't work as intended, i.e., it has bugs, then the design can be corrected and the device can be reprogrammed easily. However, most modern circuits are complex, and it is almost impossible to debug these circuits merely by observing the outputs. For that purpose, Xilinx provides hardware debug solutions (explained in Chap. 17). However, the whole process of hardware debug has its own challenges. Unless the circuit is small and simple, it is prudent to identify and correct all design issues up-front using simulation. That is, the reason simulation has become an integral part of current generation of FPGA designs. Xilinx Vivado not only provides its own simulator, but it also has most of the industry standard simulators (i.e., *Questa*, *NCSim/Incisive*, *VCS*, and *Aldec* simulators) integrated into its environment. The actual availability of the third-party simulators will depend on your license agreement with those simulators.

Vivado makes simulation very easy by providing the same framework for design and simulation. Once the design (and testbench) is set up in Vivado, it can generate scripts for seamless simulation—including for external simulators with very little to no additional change.

In this chapter we will go through the process of setting up the design for simulation, running simulation, observing the outputs, and review various tools available for debugging the design. We will also talk about Vivado's native simulator and use of C-models to speed up the simulation. As mentioned in Sect. 2.2,

S. Bandopadhyay (✉)
Xilinx, San Jose, CA, USA
e-mail: saikatb@xilinx.com

S. Churiwala (ed.), *Designing with Xilinx® FPGAs*,
DOI 10.1007/978-3-319-42438-5_11

all GUI actions get logged into *vivado.jou*, and those Tcl commands can be used to create a script, for running in batch mode for automation and regression runs. This chapter will explain some other alternative options for some of these commands. For an exhaustive list of options for these commands, you should use *<command> -help* on the Tcl console of Vivado.

11.2 Setting Up Design for Simulation

For simulating the design, you need to specify a testbench. The testbench contains HDL file(s) which provides the input to the design for simulation. It also prints and/or checks the outputs. A more complex testbench may even do white box testing by performing assertion checks on internal signals of the design.

You need to add these testbench files in addition to the design files. While adding source files from the GUI menu (*file → add sources*), Vivado provides an option to *add or create simulation source*. You need to select this option for testbench files that are not part of the design being implemented on the FPGA. These testbench files are used (along with the design files) for the purpose of simulation. Vivado determines the HDL language and variants through the file extensions. For example, files with *.sv* extension are considered as of type *SystemVerilog*.

Each simulator provides some options that you need to set appropriately. You can select the simulator and set the options by clicking on *simulation settings*. Simulation is done in three stages Compilation, Elaboration and actual Simulation. Each stage has its own set of options.

11.2.1 Compilation

Compilation is the stage where Verilog, VHDL, or System Verilog is read and a parse tree representing the model is created and stored in the design library. Most of the language-related errors are detected at this stage. *XSIM* compilation for Verilog/SystemVerilog and VHDL is performed by *xvlog* and *xvhdl*, respectively. Some of the options for xvlog/xvhdl are:

- Verilog options: To add and include paths for searching *'include*d files and also for defining Verilog macros *'define* from command line
- Generics or parameter options: To change the default parameters for top-level *module* or *entity*
- *nosort*: To prevent Vivado from trying to auto-determine the dependencies across HDL files to determine the order of parsing
- *relax*: To show some leniency toward LRM noncompliance but commonly used styles in HDL

11.2.2 Elaboration

Elaboration is the stage where parse trees are combined based on design hierarchy; parameters are resolved and a simulation kernel code corresponding to the HDL code is generated inside a design snapshot. *XSIM* elaboration is performed by *xelab* command. Some of the options for *xelab* are:

- *snapshot*: To specify the name of design snapshot meant for simulation. Default name is top-level *module/entity*.
- *debug_level*: To specify the level of debug that may be performed. It impacts the level of optimization that can be performed by the simulation engine. Values could be:
  - *typical*: For line tracing, waveform display, and deriver debugging
  - *all*: All of typical and debug of Xilinx precompiled library
  - *off*: No debugging. Provides the fastest simulation
- *relax*: To show some leniency toward LRM noncompliance but commonly used styles in HDL.
- *mt_level*: To use multi-threading for faster elaboration.
  - *auto*: Determined the level automatically, based on machine configuration.
  - *off*: No multi-threading
  - <num>: use max of <num> threads

11.2.3 Simulation

Simulation is the final stage where the simulation kernel corresponding to the design is verified and debugged by running it. In the context of Vivado's inbuilt simulator, *xsim* is the command for the actual simulation using the generated design snapshot. Some of the options for simulation are:

- *runtime*: Time for which simulation should be run, before stopping. In a typical simulation, it is the setup time, after which initial simulation stops and control is returned to Tcl shell. Simulation can continue further from Tcl shell with commands *run –all* or *run <time>*.
- *wdb*: The waveform database file that is generated from simulation. This database can be viewed by Vivado waveform viewer.
- *saif/saif_all_signals*: Used to generate *SAIF* file for power analysis.

There are also *additional options* under *compilation/elaboration* and *simulation* tabs which can be used for passing any options to the parser (*xvlog/xvhdl*), elaborator (*xelab*), or simulation engine (*xsim*).

11.3 Simulation and Observing Results

11.3.1 Simulation of Behavioral/RTL Model

Initially (before synthesis) only RTL design is available, and simulation can be performed on it via selecting *run simulation* from Flow Navigator window and further selecting *run behavioral simulation*. This is the fastest simulation and any issue found at this stage is the easiest to fix. After synthesis and implementation, the *run simulation* will also let you run post-synthesis functional/timing simulation and post-implementation functional/timing simulations, respectively. These simulations are more accurate but considerably slower.

11.3.2 Simulation Steps

On running simulation, Vivado internally calls *launch_simulation* command to run the simulation and displays the initial result. *launch_simulation* is the command for not just Vivado simulator but also for other integrated simulators. To select the appropriate simulator, set the property *TARGET_SIMULATOR* to one of *XSIM*, *ModelSim*, *IES*, or *VCS*. The default value is XSIM.

set_property TARGET_SIMULATOR <name>

launch_simulation script does the following:

- Determines design sources, including files
- Determines the order of parsing (if requested)
- Compiles all Verilog and System Verilog files with *xvlog*
- Compiles all the VHDL files with *xvhdl*
- Elaborates the design into simulation snapshot using *xelab* command
- Opens up *design scope* window, *objects* window, and *waveform* window to monitor the simulation
- Runs simulation on the snapshot using *xsim* command for a pre-specified simulation time
- Gives control back to Tcl shell for further simulation commands or for inspection of design or output

Some of the options for *launch_simulation* that might be of interest to you are:

- *step*: Fine control of simulation stage to perform. Values are *compile*, *elaborate*, *simulate*, and *all* (default is *all*).
- *scripts_only*: Only generate the simulation scripts; don't actually execute the scripts.
- *noclean_dir*: After simulation run, don't clean up the directory.

Any error in compilation or elaboration of the HDL files will be reported in the *messages* as well as *log* tab at the bottom of Vivado. The error messages in the message tab has hyperlink to the source for speeding development. Output of simulation can be observed in log window of Vivado.

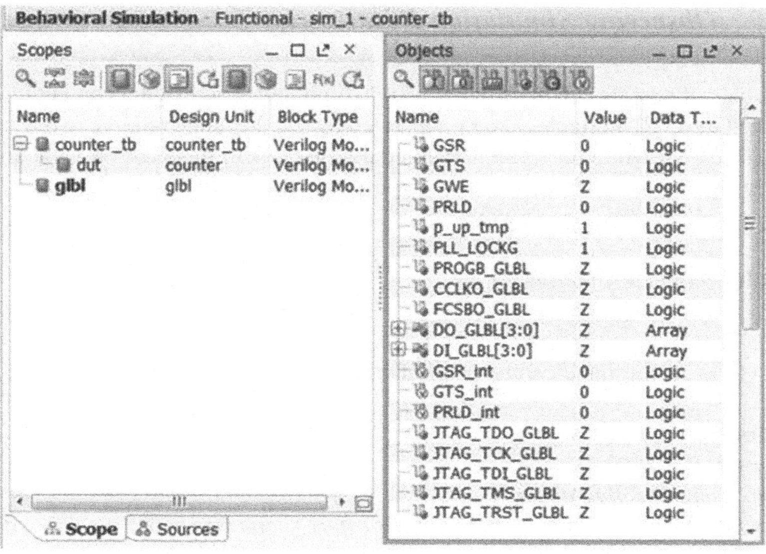

Fig. 11.1 Objects window

Scope can be browsed on the *scope* window. On selecting a scope, all the signals in the scope are displayed in the *objects* window as in Fig. 11.1. Once the scope is changed, the object window will start displaying signals of the new scope. This *objects* window displays the current value of the signal. Vivado picks the default radix to display the value. This radix can be customized by the pop-up menu on right click over the signal. To see previous values at specific time, waveform viewer can be used. This is explained in Sect. 11.3.6.

Some of the Tcl commands related to *scope* are:

- *current_scope* without any argument: Returns the name of the current scope.
- *current_scope <name>*: The scope is changed to the specified name.
- *get_scopes:* Lists all the child scopes of the current scope.
- *report_scopes* Describes all the child scopes of the current scope.

The following is an example transcript:

```
current_scope /counter_tb
/counter_tb
get_scopes
/counter_tb/dut /counter_tb/Initial28_0 /counter_tb/Always35_1 /
counter_tb/Monitor32_6
report_scopes
      Verilog Instance: {dut}
      Verilog Process: {Initial28_0}
      Verilog Process: {Always35_1}
      Verilog Process: {Monitor32_6}
```

11.3.3 Observing Simulation Results with Tcl

The values of signals can be printed by Tcl command. Each signal can be uniquely identified by a full hierarchical path separated by / or the name relative to the current scope. The following Tcl session prints the value (*28* in this case) for the signal / counter_tb/dut/i1/r3. If the current scope is set to `/counter_tb/dut`, the same signal can also be accessed by the name *i1/r3*.

get_value */counter_tb/dut/i1/r3*
28

The type of the signal can be queried with the command *describe*. That is:

describe */counter_tb/dut/i1/r3*
Port(OUT): {r3[7:0]}
Path: {/counter_tb/dut/i1/r3}
Location:{File "C:/AXI_tut/project_3/project_3.srcs/sources_1/new/counter.sv" Line 26}

To print all the signal values in the current scope, Tcl command *report_values* is used.

report_values

Declared: {count[7:0]}	*Verilog 28*
Variable: {reset}	*Verilog 0*
Variable: {clock}	*Verilog 1*

11.3.4 Timing Simulation

Xilinx maintains libraries with and without timing information. Library without timing can be used for faster verification of functionality. However, if you want to also consider the individual gate and wire delays, you should use *post-synthesis timing simulation*. In this mode, the simulators will also flag if any of the timing checks as they are violated during simulation. Post-implementation timing simulation uses. SDF files generated from Vivado to model more accurate wire delays and timing checks.

11.3.5 Controlling Simulation from Tcl

Simulation in Vivado will run for time duration specified in simulation options and will stop for further commands. Simulation can be continued with the Vivado command *run*. *run* runs the simulation further from the currently stopped time.

run 100 #runs simulation for 100 ns (ns is default time unit for simulation)
run 100 us #runs simulation for 100 micro second (timeunits are ms, us, ns, fs)
run –all #runs simulation till there are no more process in the design

Current simulation time can be observed with the command Vivado Tcl command *current_time*. To redo the simulation without the overhead of re-compilation, there is the Vivado Tcl command *reset_simulation*. This resets the simulation time to *0* and cleans up any files or data generated during simulation. If you are debugging and want to preserve *breakpoints* and *conditions*, you will need to use the command *restart* instead of *reset_simulation*.

11.3.6 Waveform Window

During simulation, Vivado generates a waveform database and displays it in the waveform window. When simulation is done for the first time, Vivado automatically displays all the signals at the top level of the design. You can add signals to waveform window by dragging signals from the *objects* window to the *waveform* window. Or, you can use the Vivado Tcl command *add_wave* with hierarchical or relative signal name. You can customize the waveform being added through the use of the following switches to add_wave:

- *radix*: To set the radix for displaying the values. Valid radix types are bin, oct, hex, dec, unsigned or ascii.
- *after_wave/before_wave*: To customize the placement of the wave. By default, the new waveform is added at the bottom of the existing waveforms.
- *color*: To set the color of the waveform, which can be a standard color name or a string of the form ##RRGGBB.
- *r*: Used to add all signals under the specified scope.

If you have customized your waveform, you can save the customization as waveform configuration, to be loaded during future simulation of the same design. To save a waveform configuration, select the waveform and press *Ctrl-S*. The configuration gets saved as a **.wcfg* file. It is possible to save multiple waveform configurations into separate wcfg files. For restoring a stored waveform configuration, select *file→open waveform configuration* from the menu and select the **.wcfg* file.

Waveform viewer also has an ability to display the data in analog form, as shown in Fig. 11.2. It can be very useful in visualizing signal processing data. To see analog wave, right-click on the signal and select *waveform style* as *analog*.

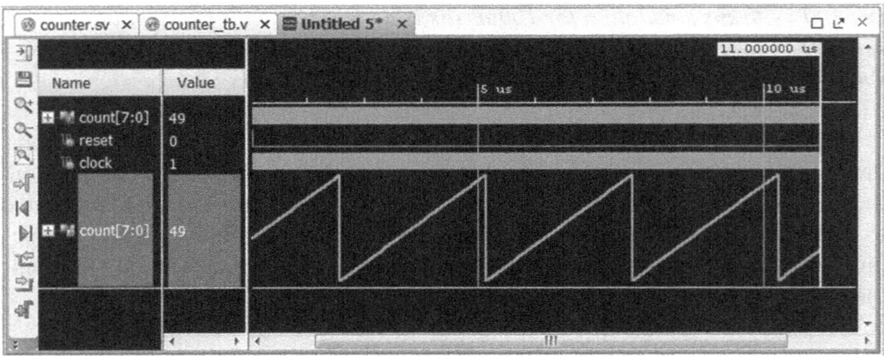

Fig. 11.2 Analog waveform

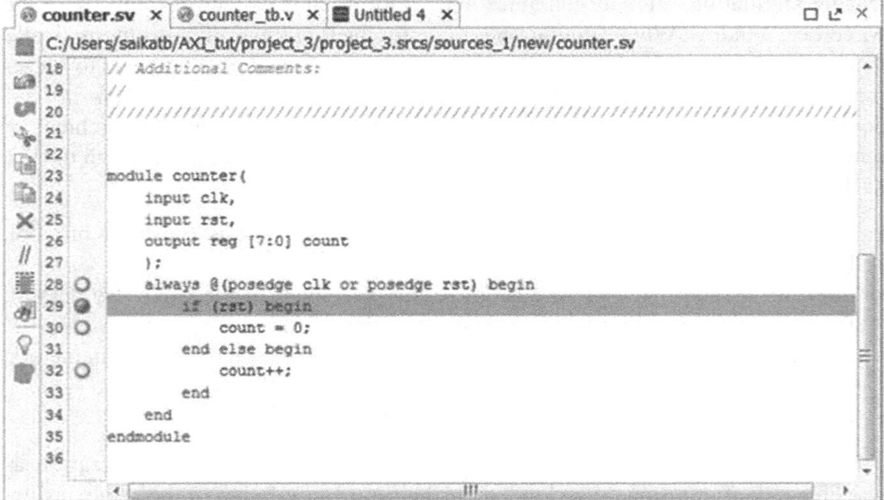

Fig. 11.3 Source window with breakpoints

11.4 Debugging

Vivado IDE has easy and intuitive ways of debugging. The first step is to analyze the waveform and/or the log to get to the simulation time, where the bug first manifests itself. Once that is identified, you should take the simulator to the specific simulation time with *run* (explained in Sect. 11.3.5) command. Tracing and HDL code debugging are done further to isolate and identify the bug.

11.4.1 Enabling Tracing

Tracing refers to ability to follow (trace) the flow of simulation on your HDL code. Vivado simulator has two tracing commands: *ptrace* for process tracing and *ltrace* for line tracing. Use *ptrace on* (or, *off*) to turn on (or, off) process tracing. An example session with process tracing would look like as below:

run *5*
INFO: /counter_tb/Always35_1
INFO:
C:/Users/ saikatb/AXI_tut/project_3/project_3.srcs/sim_1/new/counter_tb.v:35
INFO: /counter_tb/Forked24_7
INFO:
C:/Users/saikatb/AXI_tut/project_3/project_3.srcs/sources_1/new/counter.sv:24
INFO: /counter_tb/Monitor32_6
INFO:
C:/Users/saikatb/AXI_tut/project_3/project_3.srcs/sim_1/new/counter_tb.v:32
INFO: /counter_tb/Always35_1
INFO:
C:/Users/saikatb/AXI_tut/project_3/project_3.srcs/sim_1/new/counter_tb.v:35
4055: clock:0 reset:0 ==> count:10010010

Similarly line tracing can be turned on or off with the command *ltrace on|off*. An example session with line tracing would look like as below:

run *5*
4060 ns
"C:/Users/saikatb/AXI_tut/project_3/project_3.srcs/sim_1/new/counter_tb.v":36
4060 ns
"C:/Users/saikatb/AXI_tut/project_3/project_3.srcs/sim_1/new/counter_tb.v":35
4060 ns
"C:/Users/saikatb/AXI_tut/project_3/project_3.srcs/sim_1/new/counter_tb.v":36
4060 ns
"C:/Users/saikatb/AXI_tut/project_3/project_3.srcs/sources_1/new/counter.sv":29
4060 ns
"C:/Users/saikatb/AXI_tut/project_3/project_3.srcs/sources_1/new/counter.sv":32
4060 ns
"C:/Users/saikatb/AXI_tut/project_3/project_3.srcs/sources_1/new/counter.sv":28
4060: clock:1 reset:0 ==> count:10010011

Since there can be a lot of information generated by *ltrace* or *ptrace*, it might be more effective to use tracing with condition (explained in Sect. 11.4.3).

11.4.2 Breakpoint

HDL source has executable lines on which breakpoint can be applied. Breakpoint can be set from Vivado GUI or from Tcl Console. In Vivado GUI, the breakable lines have empty circles in front of them, as shown in Fig. 11.3. Clicking on this circle will add the breakpoint. The breakpoints on the GUI are toggle switch. So clicking them again will remove the breakpoint.

The Tcl command to add breakpoint is:

add_bp counter.sv 29

Running the simulator now will automatically stop on hitting this line. The signal values can be observed by hovering the mouse over the signals of interest. Values can also be checked from Tcl shell with *get_value* or *report_value* (described in Sect. 11.3.3) command. Alternately current values can also be observed in the objects window by selecting the appropriate scope.

All the currently active breakpoints can be listed by the command *report_bps*. The report of active breakpoints will look like:

report_bps
bp2:
"C:/Users/saikatb/AXI_tut/project_3/project_3.srcs/sim_1/new/counter_tb.v":32
bp3:
"C:/Users/saikatb/AXI_tut/project_3/project_3.srcs/sim_1/new/counter_tb.v":30
bp4:
"C:/Users/saikatb/AXI_tut/project_3/project_3.srcs/sources_1/new/counter.sv":29
bp5:
"C:/Users/saikatb/AXI_tut/project_3/project_3.srcs/sources_1/new/counter.sv":32

Tcl command *remove_bp* can be used to remove a breakpoint by either specifying the breakpoint id or file and line number. So the following two commands are equivalent:

remove_bp bp5
remove_bp –file "C:/AXI_tut/project_3/project_3.srcs/sources_1/new/counter.sv" –line 32

To remove all the breakpoints, you can use the option *–all*.

11.4.3 Conditions

Condition is a very powerful debugging concept. It permits an action to be associated with a Boolean expression turning true. Whenever the condition is met, the command associated with the condition is executed. You can make use of this capability through *add_condition* from Tcl command. The command associated with condition can be any valid simulator command. Thus *add_condition* can be a powerful tool to do white box testing of design without modifying it. For example, to break simulation if a non-zero data is present with reset, use the following command:

add_condition –name ignoredData {reset == 1 && data !=0 } stop

Here *stop* is the command that gets executed when (*reset==1 && data !=0*) becomes true. On stopping, the other related signal values can be inspected for debugging. Once inspected simulation can further continue with *run* command.

report_conditions reports all the condition objects that are active. *remove_condition* just like *remove_bp* can be used to remove one specific or all conditions.

11.4.4 Changing Values of Signals

For debugging, you may sometimes want to see the impact of a changed value of signal without changing the design and recompiling it. Vivado simulator provides with two ways to do that from the Vivado Tcl. They are setting and forcing values.

11.4.4.1 Setting Value

You can use *set_value* to update a signal or reg immediately. It however permits the value to be changed with future signal update events. Example use:

set_value -radix bin /test/bench_VStatus_pad_0_i[7:0] 1110100101

11.4.4.2 Forcing Value

Forcing is similar to setting value, except it can be done only for signals, and once the value is forced, it cannot be changed till the force is on. You can force a signal to the desired value through *add_force*. Similarly the force can be removed by the command *remove_force*.

You can also use GUI to *force* a signal to a constant value and to remove the *force*. Right-click on the signal in the *object* window and select *force constant* to force a signal to a constant value. Similarly select *remove_force* from the menu to remove any *force* on the signal. You can also *force* a toggling value to a signal by selecting *force clock* and filling up the pop-up window with toggle values, start time(offset), duty cycle, and period.

11.5 Combining C with HDL Using DPI

With SystemVerilog, you can write a part of design in *C* or *C++* and use it from SystemVerilog. SystemVerilog can import *C* functions via *import* command. Once a *C* function is imported, it can be called as a regular SystemVerilog *function* or *task* (depending on the *import* command syntax). Similarly SystemVerilog *function* or *task* can be exported to *C* side. These exported SystemVerilog functions (or tasks) can be called from C functions. Exported SystemVerilog tasks can be used to mimic *#delay* or *@wait* from *C* functions.

For simulating SystemVerilog with *C* code, it is important to make sure that they are compatible. *xelab* has an option *–dpiheader*. This generates a *C* header file for the imported and exported functions. The *C* function definition prototype must match with this generated header for successful linking of functions.

Use the Tcl command *xsc* to compile the *C* files and then link with *xelab*. Only simple scalar types are permitted as function return for imported or exported *DPI* functions. The permitted data types that can be passed between C and SystemVerilog are mentioned in Table 11.1.

Let us take an example to elaborate. A SystemVerilog file *hdl.sv* (with the design top name *TESTBENCH*) calls a *C* function defined in file *helper.c*. The prototype for the function as defined in *hdl.sv* is:

Table 11.1 Permitted data-types between SystemVerilog and C

SystemVerilog	Matching C type
bit	SVBit
logic or reg	SVLogic
int	int
byte	char
short	short
long	long
chandle	void*
packed array of bits	SVBitVector
packed array of logic	SVLogicVector
unpacked struct	struct

```
typedef struct {
   bit a;
   integer b, c;
} stType;

typedef struct {
   reg r;
   stType st;
} stType1;
import "DPI-C" function int func(input stType1 in, input stType1 in1);
```

To generate the equivalent *C* prototype for function, you need to elaborate the design using *xelab* with the additional command line option *–dpiheader*. This will generate a header file with the name *dpi.h*. For this case, the *dpi.h* will contain the *C* prototype of the equivalent function.

```
#include "svdpi.h"
typedef struct {
   svBit a;
   svLogicVecVal b[SV_PACKED_DATA_NELEMS(32)];
   svLogicVecVal c[SV_PACKED_DATA_NELEMS(32)];
} stType;

typedef struct {
   svLogic r;
   stType st;
} stType1;

/* Imported (by SV) function */
DPI_LINKER_DECL DPI_DLLESPEC int func( const stType1* in , const stType1*
in1);
```

SV_PACKED_DATA_NELEMS() is defined in *svdpi.h* which is included. *SV_PACKED_DATA_NELEMS* rounds the bits into chunks of 32 bits needed to hold. So number from *1* to *32* will become *1*, *33–64* will become 2, and so on. The *C* code needs to have the same prototype for *function*. The *C* code can be compiled into a dynamic library *dpi.so* with the command *xsc* as:

```
xsc helper.c
```

The next step is to create the simulation kernel with *xelab* command where the name of the *dpi* library is specified.

```
xelab TESTBENCH –snapshot SIM1 –sv_lib dpi
```

Once the kernel is created, the simulation can be run using *xsim* command. This will open up a Tcl shell that takes all the simulation commands.

xsim SIM1

11.6 Generating SAIF File for Power Estimation

For power analysis, usually functional simulation is performed and *Switching Activity Interchange Format* (*SAIF*) file is generated. This *SAIF* file is input to the power analysis tools. During simulation a *SAIF* file can be opened with Tcl command *open_saif <SAIF_file_name>*. Only one SAIF file can be opened at a time during simulation. Simulator needs to be then instructed to log signals into SAIF file, which is done via *log_saif <signal>+*. To log all the signals in the current scope, *log_saif* can be used with *get_objects* as:

log_saif [get_objects]

Once logging is done for SAIF, it can be closed with Vivado Tcl command *close_saif*.

Chapter 12
Clocking

John Blaine

12.1 Clocking in FPGA Designs

FPGAs are designed to be used with synchronous design techniques. As such, understanding clocking structures and their capabilities is vital to be able to realize a design. Poor understanding will create designs that are unreliable and difficult to meet timing, while good understanding will create reliable designs and allow you to focus on resolving non-clocking issues.

FPGA clocking is not a difficult subject to understand. Wherever you face a design decision, opt to prioritize clocking and keep the clocking as simple as possible. This simple rule will guide you well. Often decisions taken that do not give optimal clocking performance will result in delays to the project, board respins, etc.

FPGAs provide low skew clock routing. These are high load distribution networks. The network is fully buffered by design. It does not reduce in performance as you increase the load. One key progression in UltraScale FPGAs is to provide more clocking flexibility when compared to older FPGAs. There are many more available networks to use now.

Additionally, FPGAs provide PLLs/MMCMs that allow you to do frequency synthesis and phase shifting. These attributes allow you to interface to external components and generate internal clocks of almost any frequency up to the maximum operating range of the FPGA. This allows for efficient FPGA design as you can easily change the frequency at which the design operates to be optimal for the given FPGA and part of the design.

J. Blaine (✉)
Xilinx, London, UK
e-mail: john.blaine@xilinx.com

S. Churiwala (ed.), *Designing with Xilinx® FPGAs*,
DOI 10.1007/978-3-319-42438-5_12

12.2 Choice of Clock Frequency

A typical FPGA design has many clock networks, as shown in Fig. 12.1, because each of the following may have its own network:

- Each source synchronous interface coming into or leaving the FPGA
- Each transceiver interface
- Internal system FPGA clock network
- Low-speed clocking networks for control like high fanout processor control via an AXI-Lite interface, external flash clocking
- Optional internal fast clock networks for conducting DSP operations

Most designs do not run at any single clock frequency. Design frequencies are normally dictated by:

- Bandwidth of incoming data
- Bandwidth of outgoing data
- Resource consumed by a particular function

The first two points are typically decided by the system. However, the third point is a design decision, in the sense that there might be multiple combinations of freq vs. utilization that would be possible. Generating different frequency clocks is easy in a FPGA. Running something faster will usually save resource. So, you can change frequency to save FPGA resource like DSP slices.

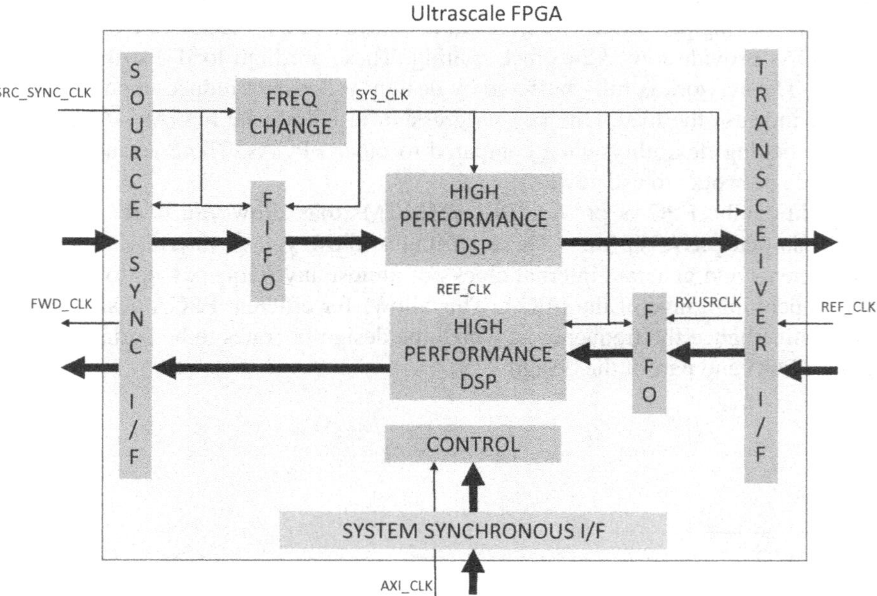

Fig. 12.1 A high level look at a typical clock network

Wireless radio designs, for example, have parts that run at sweet spot frequency of 491 MHz. Usually it is only the DSP portions that run at this performance. This includes filters, power monitors, DPD, and crest factor reduction. The designs have characteristics such as:

- Low load control paths.
- Point-to-point data paths.
- Design can be pipelined without issue.
- Data paths are typically small around 32 bits.

Wired designs tend to have a lot of switching and wide data paths. Data paths can be 512/1024/2048 bits. These large data paths represent a challenge to the FPGA design software. You can help here by selecting a frequency that balances the difficulty and data path width. These designs tend to operate in the region of 300–350 MHz. For UltraScale+, there could be benefit in doubling the frequency to something like 600 MHz and halving the data width. Smaller data widths are easier to route for the FPGA software tools.

For other types of design, you should consider data path sizes, high fanout non-clock nets, and logic levels required. These are the typical factors that influence *Fmax*. Of course, faster device families and speed grades move the window.

You should also consider productivity against the cost saving of running faster. It is important to choose the right performance without impacting your productivity level. For example, closing timing at 400 MHz may take a few extra weeks compared to closing timing at 200 MHz.

12.3 Number of Clocks

UltraScale provides capability to use up to 24 truly global clocks. Usually designs require something under 12 truly global lines. The unused networks can be broken down into many smaller clock networks. This can give hundreds of smaller clock networks. In practice, there is one smaller clock per interface, and you can use the additional remaining clocking resources for non-clock (but, high fanout signals) routing like clock enables or resets.

Vivado will handle up to 24 clocks without issue. Once you go over 24 clocks, your intervention could be required. For example, consider a design with 12 global clocks and 36 interface clocks. The 12 global clocks could be Vivado placed, and 36 regional clocks might require some user floor planning to ensure that there are no overlaps where you might exceed 24 clocks in a region.

It is possible to have many local clock networks. These are where the clock is routed on standard FPGA routing. These networks should be kept to a minimum number of loads. The recommendation would be to keep everything inside a single slice or two slices, and this would mean under 32 loads. More than this is possible but you should avoid.

12.4 A Typical Clock Network

A typical clock network (shown in Fig. 12.2) in a FPGA starts with a pin that is fed by an external oscillator. If a frequency modification is required, you should feed the incoming clock to a MMCM/PLL and then into a global clock network via a BUFG. From this point it can access clock pins of basic logic elements like flip-flops and RAMs. There are many variants of this. This chapter explains your choices depending on your requirements and optimal implementation for UltraScale FPGAs.

12.4.1 Clocks Entering a FPGA

There are two primary places where a clock source will enter a FPGA. The first is a global clock IO or GCIO. In UltraScale, there are four P/N pairs per clock region. For single-ended clock, connect on the P side. From these inputs there is a direct connection to a PLL and MMCM or BUFGs.

The other main entry points are gigabit transceivers. Each quad has two clocks that can enter the FPGA. These clocks can be used in the fabric as is or can be used with some modification of frequency. Access to MMCMs can be made through BUFGs. In Zynq MPSoC, there is a third source of clocks and that is the processing subsystem. These can pass clocks to the FPGA fabric.

12.4.2 Generating Clocks with Different Frequencies

Designs typically require many different clocks of different frequencies. FPGAs provide the facility to generate clocks of different frequency and phases using MMCMs and PLLs. PLLs can be considered as MMCMs with reduced features. Each MMCM/PLL can generate multiple output clocks at different frequencies and/or phases, over a wide range of frequencies.

MMCMs/PLLs are usually driven by (a) clocks that come from external oscillators, (b) source synchronous IO interface clocks, or (c) other internally generated clocks.

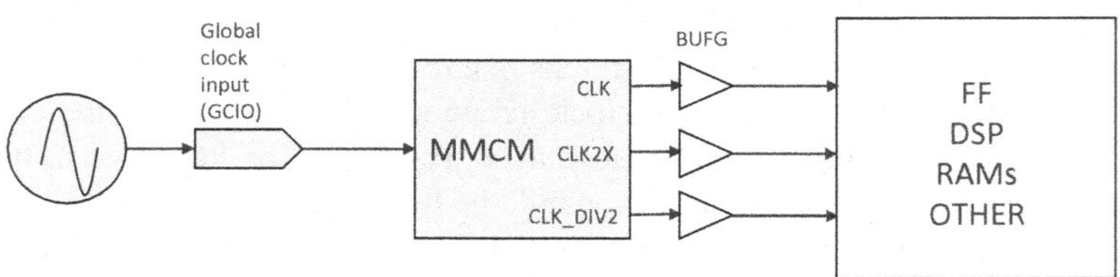

Fig. 12.2 A typical clock network

Usually you need to ensure that clocks supplied to these components should not stop. If they do, you need to reset these components. It is generally advisable to reset the components before using them.

Use the *Clocking Wizard* IP to make use of MMCMs/PLLs. This IP is part of Vivado *IP Catalog*, explained in Chap. 3.

Zynq MPSoC components also have their own PLLs to change the frequency of input clocks to the subsystem. Additionally frequency change can be achieved within a gigabit transceiver, and clock division can be done in BUFGCE_DIVs and BUFG_GTs which are described next. Detailed analysis of Zynq and transceiver clocking is not covered here.

12.4.3 Accessing Global Routing

Clock buffers BUFGCE, BUFGCE_DIV, BUFG_GT, and BUFGCTRL instruct Vivado to use the special clock routing resources. There are no special requirements to come off the global routing. Collectively these are called BUFG*. There is a BUFG primitive, which is inherently a BUFGCE, but without using the *enable* function. Vivado can infer BUFG automatically or you can instantiate them. Alternatively, IP can include them via instantiation. Only BUFG buffers can be inferred. Other types of clock buffers have to be instantiated in your HDL or in IP.

The following buffers exist in clock regions that contain IOs:

- BUFGCE—It offers an enable/disable switch. There are 24 per region. This is the base primitive.
- BUFGCE_DIV—This is similar to the above but also can divide the clock frequency. There are 4 per clock region.
- BUFGCTRL—This offers muxing capability. This is required for clock switching or multiplexing. There are 8 per clock region.

The following buffers exist in clock regions that contain GTs:

- BUFG_GT—The input has access from any of the transceiver clock sources. These have dynamic divide capability. There are 24 per region.

The access to each of the BUFGCEs in each region is independent. Each of the 24 buffers can be accessed on the input side from any MMCM/PLL output, internal FPGA resource, or IO. However, the output side will drive a particular clock track. For BUFGCE_DIV and BUFGCTRL, the input sides are shared with BUFGCEs, but the output side is flexible. Figure 12.3 shows how BUFGCE uses different clocking tracks.

The complex connections of these buffers mean that generally Vivado should decide the locations of the buffers. You can place them at the clock region level through the command below, but Vivado will determine the track numbers.

set_property CLOCK_REGION [get_cells <BUFGCE_CELL>]

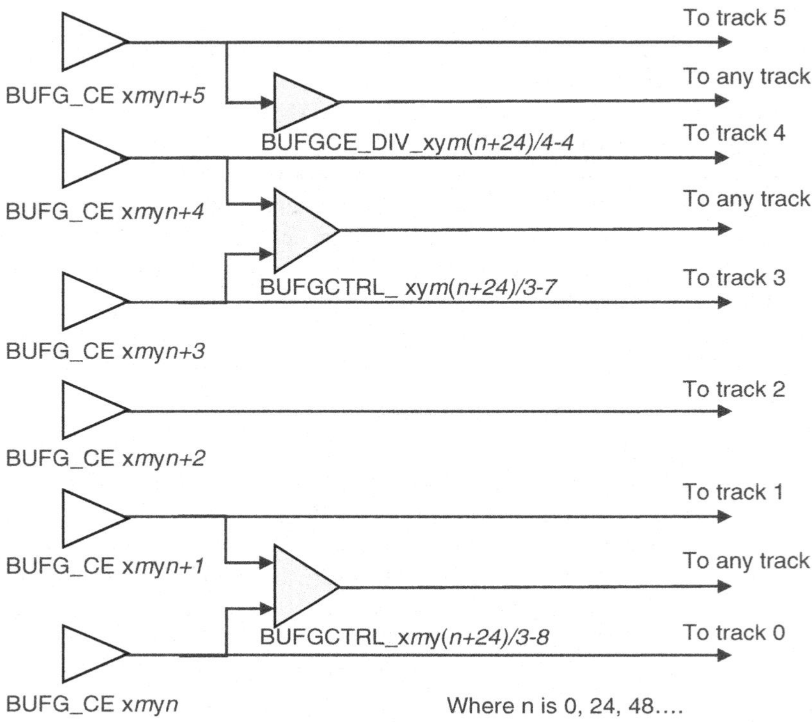

Fig. 12.3 BUFGCE output track usage by other buffers

12.4.4 Clock Routing, CLOCK_ROOT, and Clock Distribution

Due to additional clocking resources, UltraScale has introduced new terms. From the output of one of the BUFG*, the clocks travel on clock *routing*. This is new term to describe the wires after a BUFG. Each clock region has 24 of these tracks. The point at which the clock signal transfers to *distribution* resources is termed the *CLOCK_ROOT*. Distribution resources will carry the signal to flip-flop clock pins and other endpoints. Each clock region has 24 of these too. They are mutually exclusive in each clock region, except that both will be used in the CLOCK_ROOT region. Figure 12.4 explains the terminology.

Clock roots can be seen after place design using the following TCL command:

report_clock_utilization –clock_roots

Vivado will approximately choose the geometric mean of the locations of the load on the clock tree to determine the CLOCK_ROOT. Balancing the clock root is important as it impacts clock tree skew which impacts timing. When interfacing to

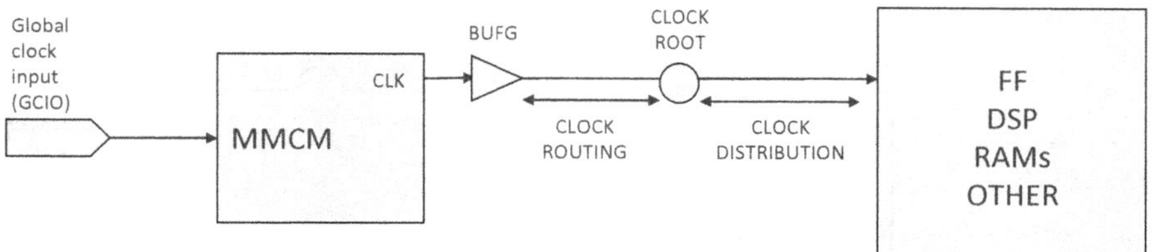

Fig. 12.4 Clocking terminology explained

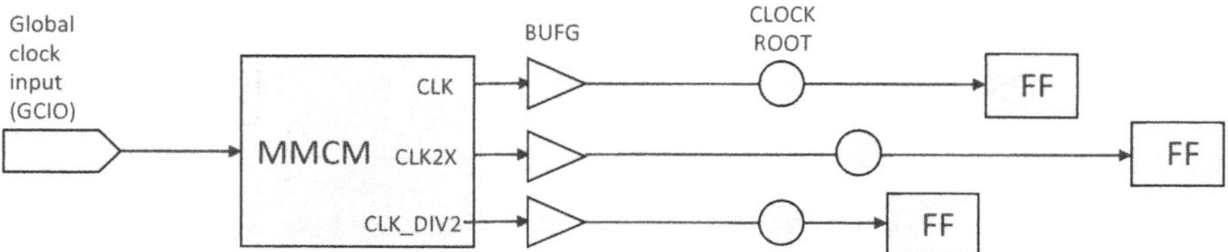

Fig. 12.5 Difference in delay due to clock root placement

a GT, CLOCK_ROOTS may be placed close to the GT in order to meet a skew requirement on transceiver clock network. This can mean that the CLOCK_ROOT is not in the geometric center. You can control the CLOCK_ROOT using the following constraint:

set _property USER_CLOCK_ROOT <clock_region> [get_nets <clock net after BUFG>]

When two or more clock networks have the same source MMCM, they can go to different CLOCK_ROOTS. This can mean that there are different delays on the clock networks, as shown in Fig. 12.5. Different delays on clock networks will translate to skew in timing which can make designs difficult to close timing.

If there are paths that are related between these clocks, you should link the two clock networks using *USER_CLOCK_GROUP* constraint.

set_property USER_CLOCK_GROUP <group_name> [get_nets [list <clock after BUFG1> <clock after BUFG2>]]

This will ensure that clock paths have roughly similar delay as shown in Fig. 12.6.

CLOCK_ROOT impacts the general skew that a network has. Clock skew can be positive for setup when moving away from the clock root and negative as you move toward the clock root (Fig. 12.7). Smaller networks will have optimal CLOCK_ROOT placement with lesser skew. For this reason, it can be beneficial to break a larger clock network into two smaller ones if the interface timing can be managed, e.g., using a FIFO where the networks cross. For this to be effective, there should be no cross clocking timing.

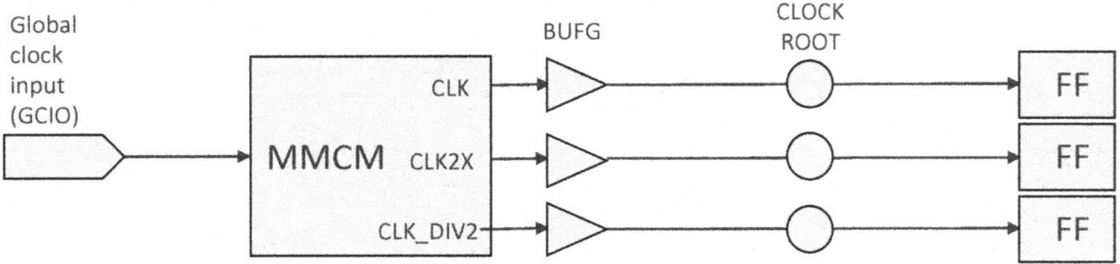

Fig. 12.6 Clock roots and length of clock tracks aligned through USER_CLOCK_GROUP

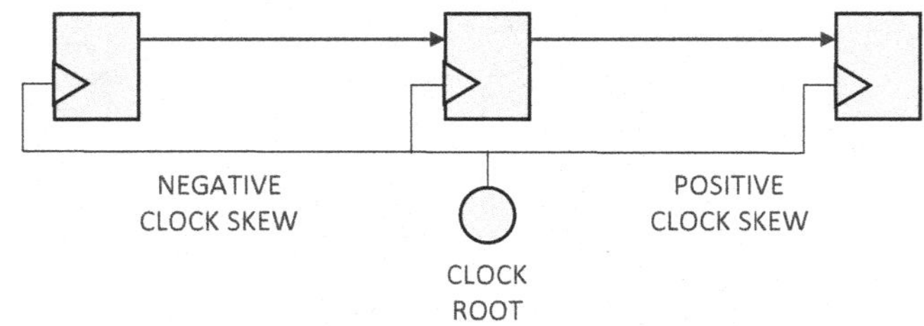

Fig. 12.7 Clock tree skew impacted by CLOCK_ROOT

12.5 Optimizing Clock Networks to Improve Internal Timing

Vivado models the most pessimistic timing. That means for setup analysis, the source path will have maximum delay, while destination path will have minimal delays. These conflicting models can significantly reduce timing budget but are required to generate a design that works in hardware in all devices.

12.5.1 Clock Pessimism Removal and the Common Node

Clock pessimism removal is compensation in a timing report for the common segments in the clock paths. It is not possible to have both the best and worst case occurring at the same time, on any given path segment. Vivado compensates for the unnecessary pessimism due to delay differential on this common segment. The point at which the source and destination clocks diverge is termed the *common node*, as shown in Fig. 12.8. Having the *common node* as close as possible to both the source and destination will improve timing margin significantly as significant portion will be common, from where pessimism would be compensated.

You can influence *common node* during early clocking decisions, such as clock frequency to use. If you choose a common clock, the *common node* will be somewhere after the BUFG. If you choose different clocks, then the *common node* will

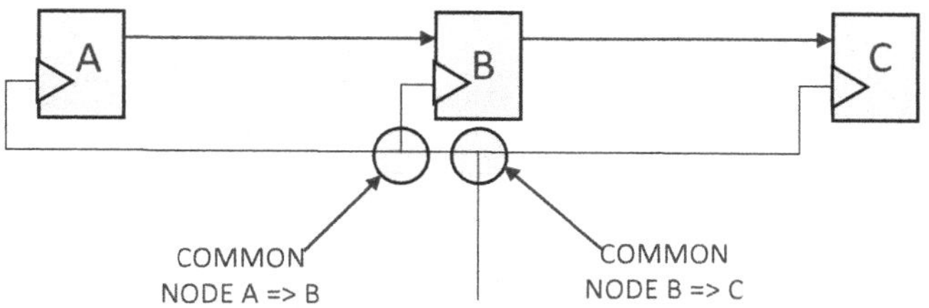

Fig. 12.8 Common node

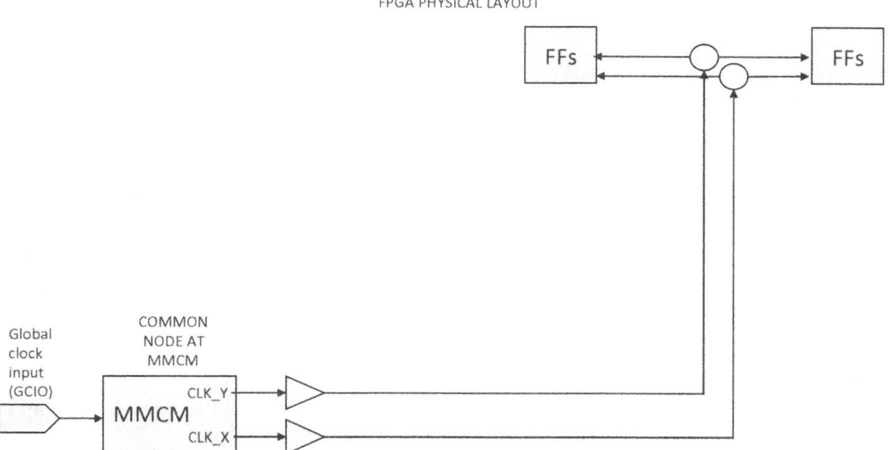

Fig. 12.9 Example circuit with a poor common node

be at the MMCM. If you opt for different clocks, then consider using asynchronous design techniques to cross the clock boundaries to improve timing. Using, for example, a FIFO will mean that timing can be relaxed at this point.

Minor *common node* influences, such as different slices, are largely controlled by the placer. However, if manually creating a floor plan, keep to clock region boundaries for optimal common nodes. Horizontal floor plan shapes are more optimal than vertical ones but, sometimes, some other considerations (e.g., data flow through DSP chain) may cause you to prefer vertical shapes.

12.5.2 Optimizing Common Node for Synchronous Cross Domain Crossings

In UltraScale, the total length of a typical clock network is made up of MMCM → BUFG → CLOCK_ROOT → LOADs. When crossing between two different clocks, the common node will be at the MMCM/PLL, as shown in Fig. 12.9.

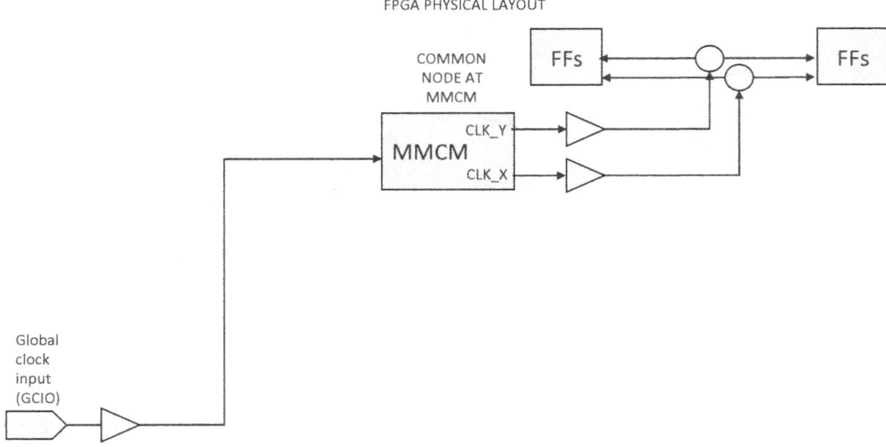

Fig. 12.10 Optimization of common node

In the case where the MMCM is not required for IO interfaces, it can be optimal to move the MMCM/BUFGs as close to *CLOCK_ROOT* as possible, as shown in Fig. 12.10. This has the effect of improving the common node, and hence greater clock pessimism removal is seen. An initial run through the tools is required to achieve this optimization.

In order to do this optimization, you must instruct Vivado as the default is to have the MMCM next to the global clock input, through the following:

1. LOCK the MMCM to the clock region close to the CLOCK_ROOT.
2. Insert a BUFGCE between the IO and the MMCM.

12.5.3 Phase Error

When two clocks come out of the same MMCM and there are timing paths between them, a value for phase error is added to both setup and hold times. This value is 120 ps for both windows. Together this creates a window of at least 240 ps that reduces timing margin. In reality, when common node compensation and hold time fixing are added, approximately 1200 ps are lost from the setup window. This should be taken into account when crossing between related clocks.

12.5.4 Internally Related Clocks Divisible by 2, 4, and 8

In a special case where the clocks are multiple of each other, use BUFGCE_DIV from a single output of MMCM/PLL. This will remove phase_error and improve the *common node*.

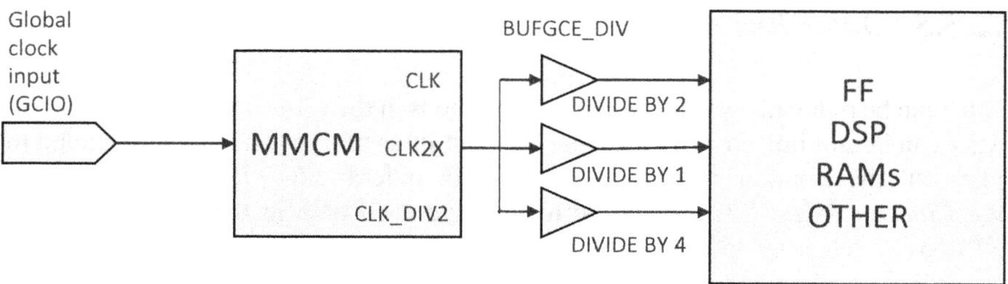

Fig. 12.11 Optimized clock network for synchronous clocks

BUFGCE_DIV primitives can divide the clock by an integer number between 1 and 8. Since there are four BUFGCE_DIVs in a region, you can derive up to four divided clocks. Consider the example circuit shown in Fig. 12.2. The clocking can be improved as shown in Fig. 12.11. The key steps are:

- Generate an MMCM with just the highest frequency output, in this case CLK2X. The *Clocking Wizard* IP should use *no buffer* as its drives selection
- Connect up 3 BUFGCE_DIV buffers in parallel
- Even for the original clock, insert a BUFGCE_DIV to divide-by 1, which helps achieve uniform delay in the clock paths

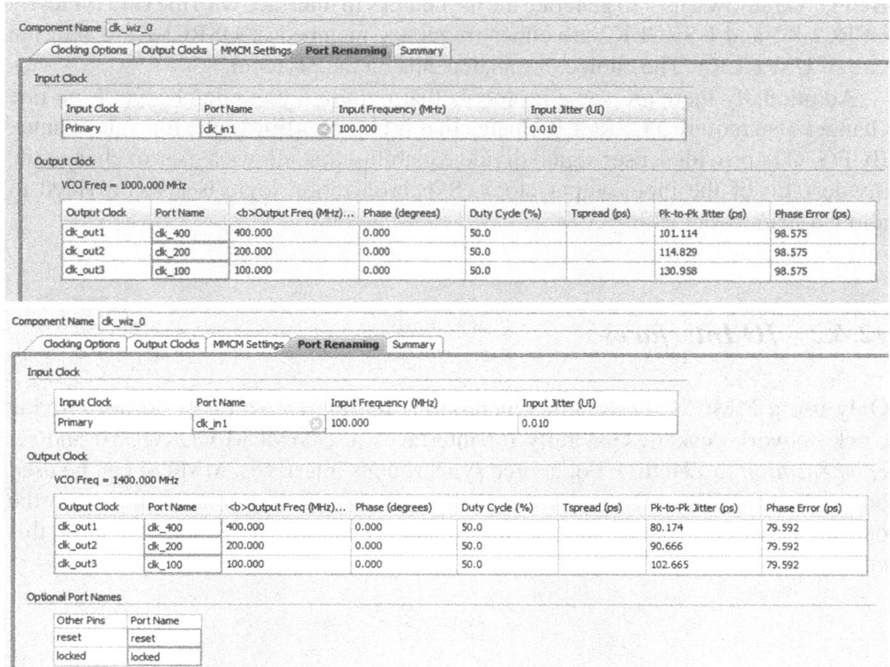

Fig. 12.12 (**a**) One combination of output jitter values. (**b**) Alternate combination of output jitter values

12.5.5 Jitter Reduction

Jitter can be reduced by selecting different options in the *Clocking Wizard* IP. Output jitter can be minimized at the expense of input jitter filtering. It is recommended to play with these options and evaluate the output under the *Port Renaming* tab inside the *Clocking Wizard*. It is possible to improve each path in the design by up to ±150 ps by selecting optimal settings. Figure 12.12a, b shows two difference values for peak-to-peak jitter for outputs, using two different settings (*balanced* and *output jitter optimized*, respectively). Be aware that change in these jitter values could impact power also, because higher frequency of VCO will result in higher power.

12.6 Optimizing Clock Networks for Interfaces

12.6.1 GT Clocking

GT clocking is generally taken care of by the IP. However, there are use cases where proprietary protocols need implementing, and in this case clocking should be understood. In UltraScale, MMCMs and PLLs are generally not required for GTs. This makes the clocking much more scalable to the GT count. Instead, dividers in BUFG_GT allow a user to generate the user clocks to interface with the GT. Typically USRCLK1 and USRCLK2 are either frequency matched or USRCLK2 is half the rate of USRCLK1. The choice of this depends on the protocol.

Additionally there are some protocols that require a line rate change. Line rate changes also require a USRCLK change that is in proportion to the line rate change. BUFG_GTs provide a user signal divide capability that allows a user to change the divide ratio of the input/output clock. Synchronization logic is also provided to allow a seamless clocking change.

12.6.2 IO Interfaces

Only use a MMCM for system synchronous IO interfaces. PLLs do not provide clock network deskew. Generally for interfaces, use MMCM CLKOUT0 and set *compensation* to ZHOLD. For source synchronous interfaces, MMCM can be used but setting for ZHOLD may adversely impact timing. You should play with the options here to establish good timing. Good constraints are mandatory for this approach.

Chapter 13
Stacked Silicon Interconnect (SSI)

Brian Philofsky

A growing trend in the semiconductor market is to gravitate toward 2.5D and 3D technologies as a way to extend and improve the growth and integration path that Moore's law has paved for more than 50 years. Xilinx has been a forerunner into this emerging technology entering into this foray in 2011 with the public introduction of the Xilinx XC7V2000T device utilizing four active die on a passive interposer creating not only the largest FPGA of the time but one of the first commercially available examples of this new technology. Since the introduction of that device, several other devices have followed and now are becoming a more mainstream means to realize large, high-performance devices to address some of the most demanding FPGA designs. Due to the sheer size and unique construction of these devices, a new approach to design should be considered in order to facilitate design entry, implementation, and closure.

13.1 SSI Terminology

With Xilinx being one of the first companies to release a 2.5D device, no established terminology for the details of the technology existed prior. So as a means to communicate this, several new terms were created by Xilinx to describe the differences in the 2.5D devices. Figure 13.1 shows a representation of SSI device.

- *Monolithic Device*: Single-die or non-SSI device
- *Super Logic Region (SLR)*: An active die in an SSI device construction.
- *Stacked Silicon Interconnect (SSI)*: The 2.5D structure utilizing multiple active SLRs attached and connected to a passive interposer

B. Philofsky (✉)
Xilinx, Longmont, CO, USA
e-mail: brian.philofsky@xilinx.com

S. Churiwala (ed.), *Designing with Xilinx® FPGAs*,
DOI 10.1007/978-3-319-42438-5_13

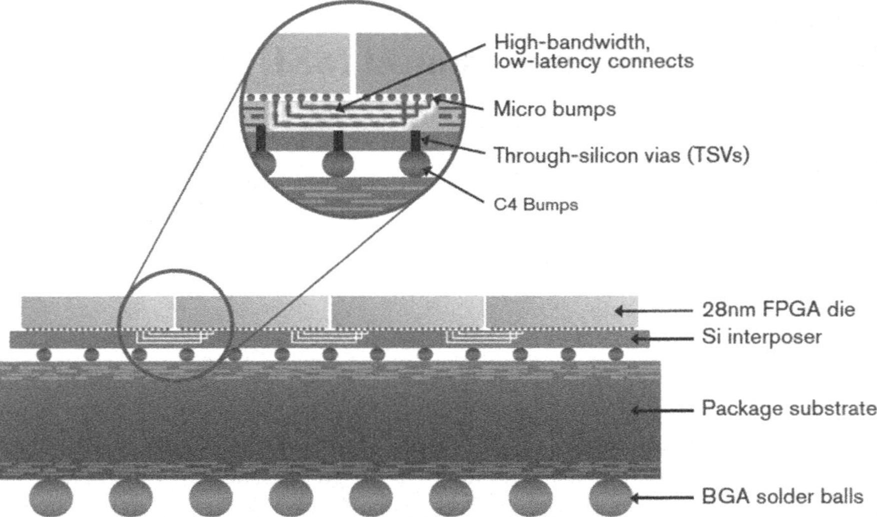

Fig. 13.1 Representation of an SSI device (*source*: Xilinx)

- *Interposer*: A passive layer in the construction of an SSI device that serves the purpose of power delivery, configuration connectivity, and connectivity between SLRs as well as connects the SLRs to the package substrate via through-silicon vias (TSVs)
- *Super Long Line (SLL)*: The active signals used to connect one SLR to an adjacent SLR in an SSI device.
- *Laguna*: The dedicated interface to traverse from one SLR to another via an SLL. This interface may or may not use a dedicated register.

The first thing that should be stated is that SSI devices do not require an all new design methodology. It is possible to target an SSI device using the same top-down method generally applied to monolithic devices with no difference in design entry, implementation, and validation. The thing that needs to be realized however is that ignoring the size and construction of the underlying device may lead to less optimal results and a longer design cycle.

13.2 Design Partitioning

One of the first SSI-specific decisions is to either chose to manually select or partition the logic to each underlying SLR in the device or to allow the tools to automatically partition the design into the separate SLRs. Vivado has the ability to take a single definition of the design and decide what portions of the logic should be placed into which SLR. The primary benefit of automatic partitioning is the obvious up-front ease-of-use benefit of not having to make such a decision, and it is very possible that you may see better out-of-the-box performance and results from automatic partitioning. Automatic partitioning can also result in higher device utilization and can potentially adjust to significant design changes more easily than manual partitioning. The drawback however is the loss of control of the design placement in the FPGA which may yield less repeatability and control during timing closure. In situations where timing closure may prove difficult, this may be a very important trade-off to consider as the added control may allow much quicker timing closure for difficult designs.

The primary design parameter that often dictates the better flow has to do with performance requirements and how much margin there is in the design to meet those requirements. For designs that wish to push the limits of the device in terms of performance or for designs in which it is desired to ensure that areas of the design that remain unchanged to have similar place and route results in future runs, manual partitioning is generally the better choice. An important thing to note is that performance limits are not always dictated by desired clock rate. For instance, for a design that has low latency or lack of pipelining, several logic levels or high fanout nets may have a much lower maximum clock rate than one that is highly pipelined. For this example, a much lower clock frequency may be pushing the performance limits of the device compared to that of a well-crafted, pipelined version operating in that same device. Following good overall design practices promotes more performance margin in the device in general and can lead to more flexibility in such design decisions. The main thing to consider is how much performance margin is expected for the design. For designs that have adequate performance margin, either method (manual floor planning or auto derived) may be suitable.

13.3 Pinout Considerations for SSI Designs

Another important up-front consideration is the pinout selection. Whether using an automatic or manual partitioning style, selection of which I/O pins are located in which SLR has a substantial impact on the associated logic placement and routing of the design. A well thought-out pinout selection will lead to good dataflow through the FPGA leading to good implementation results, providing better utilization, timing, and power. The best approach to determine a good pinout starts with examining the expected dataflow of all portions of the design and how that maps to the I/O resources of the device. All associated control signals such as clocking, enables, and resets should be considered in conjunction with this data flow.

The data path represented in Fig. 13.2 originates at SLR 0 (bottom SLR), must go to the SLR 3 (top SLR) in order to buffer the data to an external memory, and then travel back to SLR 1 to exit the device. This pinout selection has some obvious drawbacks. First off, the data path is required to traverse six SLRs to complete the data cycle. This could cause possible timing and resource issues. Also notice that the clock and reset signals that must drive all of the logic are located in SLR3. Since

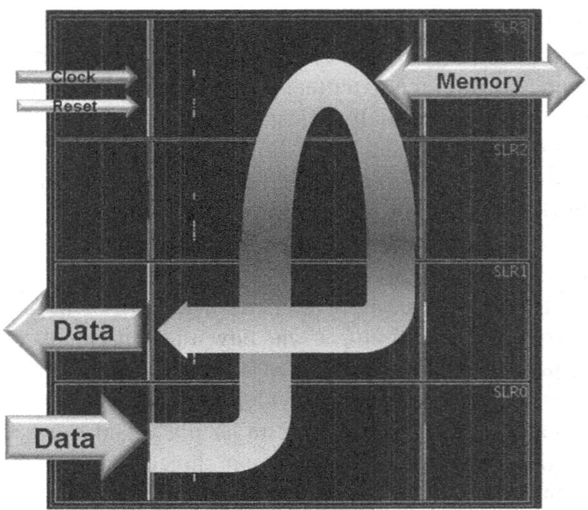

Fig. 13.2 Simple representation of clock and dataflow through an SSI device

these are high fanout signals, a better selection would be in the center SLRs so that the signals can be more evenly distributed improving the overall timing paths of both signals. Placing the reset in the center would likely reduce the overall delay from source to destination as the overall distance between these points is minimized. Moving the clock to a more centralized location will also reduce the insertion delay but, more importantly, it balances the overall clock skew as well. With the clock placed in the top SLR, the data path entering the chip to the external memory interface must travel against the clock. This increases the amount of negative clock skew for that portion of the path resulting in reduced timing margin. The best placement for the clock has to do with what portion of the design is expected to have the least timing margin and placing the clock so that it is either in the same SLR or the clock travels with the data. Having the clock travel in the same direction as data will improve setup timing margin by making use of useful (positive) skew.

Figure 13.3 shows the same data path with the memory interface moved to SLR1 and the clock and reset relocated to SLR0. The benefits of this change should be fairly evident. The overall SLR crossings reduced from 5 to 1. The high fanout reset now only needs to reach two SLRs rather than all four and is located closer to the logic it must drive. The clock is placed in SLR 0 so that the overall data path is traveling with the clock rather than against it, promoting better skew characteristics for this path. This pinout will likely consume less routing and fewer logic resources as less pipelining and resource impacting optimizations like logic replication should be necessary. Such simple pinout changes could have a dramatic impact on the overall implementation results of this design.

Often pinout decisions must weigh a balance between board layout considerations and internal dataflow optimization. Due to the size of SSI device packages, often there are numerous high-speed connections that must be routed out of a dense ball grid array that pose its own difficulties in PCB routing and power delivery. This

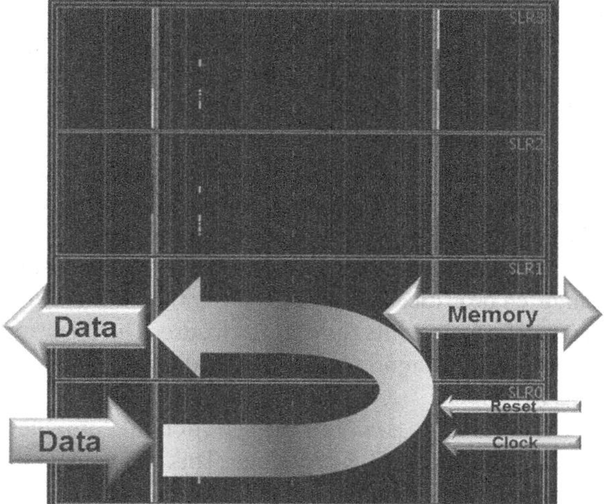

Fig. 13.3 Alternative pinout (compared to Fig. 13.2)

often leads to an iterative approach between the digital design team and the printed circuit board team trying to find the best compromise between internal dataflow and external PCB routability. Extra time and effort spent at this point can pay large dividends later in terms of easing design timing closure and reducing the overall design implementation cycle while also requiring fewer device resources and less power consumption as once pinout decisions are fixed, it is very difficult to change later.

13.4 Partitioning Considerations

For automatic partitioning, there is no difference in how you would operate the tools over targeting a traditional monolithic device. Manual partitioning however involves more up-front planning. Logical hierarchical boundaries, selected IP, division of design between different engineering teams, or combination of these factors are used to decide what logic should be physically placed in which SLR. The mechanics of manual partitioning is not all that different from floor planning, which is a common design closure technique.

13.4.1 Limit SLR Crossings

There is a relatively high but hard limit in the number of signals that can cross from one SLR to another, so ensure that any selected partition does not exceed the available SLLs for the selected device. Make an effort to understand the number of available SLLs per-SLR crossing as well as the expected SLLs required by the design as criteria in selecting a partition. The example in Sect. 13.3 showed that the number of SLR crossings is often dictated by the dataflow of the design. So, good pinout selection is important. Once the dataflow of the design is established, the mapping of the associated logic to those datapaths is often a much simpler task. After decisions as to which logic hierarchies should be mapped into which SLRs, the number of SLR crossing required for the design can be tallied and tracked. In general avoid using more than 60 % of the given SLLs in the device for any given SLR crossing. The reason behind this is twofold. Higher utilization of SLLs requires more trade-offs for placement and routing of those SLR crossing, and since the SLR crossing can often be critical paths in the design, building additional flexibility for placement can pay large benefits for timing closure. The other reason to reduce utilization is to allow future design changes and additions without concern for overutilization of SLL resources.

Report Design Analysis has a section that provides information on modules that are contributing to SLR crossings. This can help you in making decisions related to floor planning—by blocking this module to an SLR—if possible.

On the other hand, keep in mind that trying to reduce the number of SLR crossings could cause you to place more and more logic within an SLR—causing higher utilization within the SLR. Hence, it is good to maintain a good balance.

13.4.2 Limit Timing Critical Paths Across SLRs

Where possible, pipeline the data paths for the SLR crossings. If it becomes necessary to have signals that cross multiple SLRs, add additional pipeline stages accordingly to ensure adequate timing margin for such crossings. Making the decision up-front as to pipeline these interfaces often makes it much easier to understand and balance the pipeline stages compared to adding it in later stages of the design. A good practice is to use a common naming convention for registers intended for SLR interface crossing as it can help with timing analysis and floor planning later. It may also be necessary to add synthesis attributes on these pipeline registers to prevent synthesis from using shift register LUTs for those structures when multiple pipeline stages are used in a single SLR. When possible, locate control logic and the associated control signals from the data paths in the same SLR. For control logic that must span multiple SLRs, it is often best to place the logic into a centralized SLR to the partitioned logic in order to allow a more evenly distributed timing. If it is suspected that timing may still be critical, replication of the logic may become necessary in order to best manage locating the source logic to its loads. This is not much different from the process used in monolithic design. The only difference is that these decisions now have ramification into the partitioning decisions for the logic.

13.4.3 Balance Resource Usage

Resource management is important whether using automatic partitioning or manual partitioning; however, manual partitioning adds the extra task of determining a proper balance of those resources across the different SLRs. It should be planned so that any one SLR does not become too full as to negatively impact place and route results. The actual target for maximum resource usage depends on the resource type, performance requirements, and the interaction between them. For instance, a design which has a lot of margin for performance may be able to fill the SLRs to a larger capacity than one that can barely meet timing with the current requirements and design characteristics. Also, often larger blocks such as RAM blocks or DSP blocks are more difficult to get higher utilization within the SLR than more common blocks like LUTs or registers. An effective strategy for designs requiring a high percentage utilization of a particular resource is to trade off for another. For instance, if a particular design requires a high percentage of block memory, utilizing a lesser amount of LUTs and DSP often helps in the overall place and route results. Another consideration is future design growth. A good partitioning plan accounts for areas of the design that may grow so that as the design definition changes, the partitioning does not need to change to account for that.

13.5 SSI Synthesis Techniques

Once the SLR partitioning has been decided, creation of the design can commence. Design entry does not have to differ from that of a monolithic design creation. Functional verification is also no different. Synthesis and implementation may be augmented to help with this design flow. There are three possible approaches to synthesis for a manually partitioned design. In all of these synthesis approaches, the common theme is maintaining designated SLR partitions during the design flow which should maintain the intended data paths between SLRs.

13.5.1 Top-Down Synthesis

One method is to use a standard top-down approach. For the instances at SLR partition boundaries, place *KEEP_HIERARCHY* attributes. The *KEEP_HIERARCHY* attribute as the name implies instructs synthesis to retain the hierarchy in which it is applied which limits optimization at and across that hierarchy. Using this attribute prevents synthesis from moving intended logic from one hierarchy boundary to another and should retain all logic structures including additional pipelining at that boundary. Strategically placing this attribute only on the hierarchy instances that border an SLR allows synthesis to optimize across logical hierarchies contained solely within an SLR while preventing optimization of logic across designated SLRs.

13.5.2 Bottom-Up Synthesis

This approach generally consists of synthesizing each SLR or portions of SLR logic in its own separate project. This methodology by design will prevent optimization of logic across the designated boundaries but also can help facilitate team design by allowing multiple portions of the design to be implemented and verified in parallel. This method also does the best job to retain the results of areas of the design that have not changed from iteration to iteration. There are other benefits such as the ability to apply unique synthesis options for each portion of the design, and often overall runtime between iterations is reduced with this method. The drawbacks are that multiple synthesis projects must be maintained, and design coordination and assembly could become a little more difficult compared to the top-down approach.

13.5.3 OOC Synthesis

Yet another method that you could use is to synthesize the designated SLR partitions out of context with each other. This design methodology is a hybrid approach in that a single project can be maintained; however, the individual SLR partitions can be implemented independent of each other and later assembled when the top-level of the project is implemented.

13.6 SSI Implementation Flow

Once synthesis is completed, analyze the synthesis results performing utilization, DRC, and any other reports indicating the readiness for implementation. You should also perform timing analysis to ensure adequate timing margin with the unplaced design prior to starting implementation. Performing such steps is good practice for any FPGA design but is more crucial for SSI design since implementation runtime can be much longer due to the overall design size. Once you have verified synthesis results, you can run implementation or place and route. The place and route flow for SSI implementation is no different than it is for a monolithic design. It is suggested to implement the design using default options and evaluate the results. If timing and power requirements are met, then there is nothing more that is needed. Even the best planned designs may not always be successful the first time they are implemented; so if timing is not close to being met, then timing analysis should be performed with appropriate action to the design and/or synthesis settings and attributes. If the critical timing path crosses an SLR or multiple SLRs, reexamine and evaluate the partitioning. If possible, consider additional pipelining, logic restructuring, or fanout reduction. If the critical timing paths exist solely inside a single SLR, apply the same timing closure techniques (refer to Chap. 14) used for monolithic designs. If timing is close to being met, there are some strategies in the Vivado tools that apply specific algorithms for SSI devices. These SSI-specific strategies are identified with either the term SLR or SLLs in them. For instance, the strategy *Performance_ExploreSLLs* is a performance-oriented strategy impacting SLR placement and routing algorithms. Try one or more of these strategies to attempt to find additional improvement over the default implementation results. If several CPU cores are available, multiple strategies can be run in parallel on a single or multiple machines cutting down on the overall implementation runtime in using several strategies.

13.7 Examining SSI Results

Analysis of the implementation results for an SSI design is largely similar to that of a monolithic. There are no additional steps or reports to create; however, some of the existing reports will show additional information pertaining to SSI. When generating a utilization report for instance, a few additional sections appear for an SSI device that contains useful information about the implementation run. The first two SSI-specific sections contain information about the SLR crossing and clocking.

```
12. SLR Connectivity and Clocking Utilization
---------------------------------------------

+----------+-----------------+---------+-----------------+---------------+-------+-------+
|          | Total SLLs Used | (%)SLLs | BUFGs/BUFGCTRLs | BUFH/BUFHCEs  | BUFRs | MMCMs |
+----------+-----------------+---------+-----------------+---------------+-------+-------+
| SLR2     |                 |         |              13 |             0 |     0 |     3 |
| ||||||-> |            5150 |   29.80 |                 |               |       |       |
| SLR1     |                 |         |              19 |             0 |     0 |     0 |
| ||||||-> |            1786 |   10.34 |                 |               |       |       |
| SLR0     |                 |         |              12 |             0 |     0 |     1 |
+----------+-----------------+---------+-----------------+---------------+-------+-------+
| Total    |            6936 |         |              44 |             0 |     0 |     4 |
+----------+-----------------+---------+-----------------+---------------+-------+-------+

13. SLR Connectivity Matrix
---------------------------

+------+------+------+------+
|      | SLR2 | SLR1 | SLR0 |
+------+------+------+------+
| SLR2 |    0 | 5150 |  146 |
| SLR1 | 5150 |    0 | 1786 |
| SLR0 |  146 | 1786 |    0 |
+------+------+------+------+
```

This is a good indicator of how well the design is partitioned in the device. Looking at the total SLLs and usage percentage ensures that they are within the expectations and goals for the design. Analyze the clocking to ensure that it is distributed as anticipated. Within the SLR connectivity matrix section, the more interesting information is concerning the number of signals that must cross multiple SLRs, SLR0 to/from SLR2 for instance. While it is many times impossible to get that number to zero, keeping that number as low as possible is a good indicator of a well-portioned design. The remaining sections more pertain to SLR resource utilization statistics:

14. SLR CLB Logic and Dedicated Block Utilization

SLR Index	CLBs	(%)CLBs	Total LUTs	Memory LUTs	(%)Total LUTs	Registers	BRAMs	DSPs
SLR2	42765	95.54	244699	31078	68.34	373855	803.5	165
SLR1	44753	99.98	310309	39977	86.66	528056	931.5	386
SLR0	40824	91.21	216975	24647	60.59	331262	647	159
Total	128342		771983	95702		1233173	2382	710

15. SLR IO Utilization

SLR Index	Bonded IOBs	(%)IOBs	Bonded IPADs	(%)IPADs	Bonded OPADs	(%)OPADs	GTXs
SLR2	16	10.26	0	0.00	0	0.00	0
SLR1	5	3.68	0	0.00	0	0.00	0
SLR0	17	10.90	0	0.00	0	0.00	0
Total	38		0		0		0

These sections indicate whether per-SLR utilization goals are met as well showing the balance of resources across the different SLRs. You need to consider changes if a particular resource appears to be close to full utilization in a given SLR that can impact placement, timing, or future design growth.

Timing reports are another place where SSI-specific information can be found that is important to understand in terms of analyzing the results of the implementation run. When analyzing any failing path, look whether the data path crosses one or more SLRs. This is clearly denoted in that section of the report with an *SLR crossing* notation followed by the originating and destination SLR numbers. If this indicates that more than one SLR is traversed, timing will be difficult to meet for that path. If there are several logic levels or a very high fanout, net is created on an SLR crossing that may make it difficult to meet timing. If any of these situations are encountered, the more common approach to address the timing issue is to either change the logic in the failed timing path to reduce depth/fanout, repartition the logic to reduce SLR crossings, or add additional pipelining.

For high-speed logic paths that must cross SLRs, clock skew and clock uncertainty due to inter-SLR compensation should also be analyzed for improvement. Due to the size of SSI devices, clock skew can be a larger impact to timing than in smaller devices. First you want to ensure that a good clocking topology is used incorporating proper clock buffer usage and no logic exists in the clock tree as poor clock management impacts can be magnified in these larger designs. When crossing SLRs, it is generally best to not at the same time cross common clock domains as well as that can introduce additional clock skew to the timing path. Inter-SLR compensation is a calculation applied to timing paths that the source and destination exist in separate SLRs to account for uncertainties in the clocking due to PVT (process, voltage, and temperature) differences between two SLRs. Minimizing clock skew as well as the distance traveled and thus insertion delay between points crossing the SLR are ways to better manage the impact of inter-SLR compensation.

A true benefit to the manual partitioning approach is that, if done properly, any critical timing paths found after place and route are typically contained within an

SLR, and general timing closure techniques can be applied to address them. It is also often found that, once functional and timing closure is completed on an SLR, logic changes to other areas of the design have little to no impact on that portion of the design. This can be less true for general monolithic or automatically partitioned design strategies.

13.8 Divide and Conquer

There may seem to be more up-front work to design with SSI and there often is. However, having a good pinout and partition can result in less overall design cycle and fewer iterations for design closure. Effectively, a very large design is broken through a good partition into smaller more manageable pieces. Using good design practices taking into account the SSI device size and structure can result in less power, less area, and higher performance, all of which often results in fewer design iterations and a shorter overall design cycle.

Chapter 14
Timing Closure

Srinivasan Dasasathyan

14.1 Introduction to Timing Concepts

Timing closure involves modifying constraints, design, or tool flow/settings to meet timing requirements. In Vivado tool, the timing constraints are entered in *XDC* format. XDC constraints are based on the standard Synopsys Design Constraints (SDC) format.

For brevity all the constraints that Vivado supports are not explained in this chapter but only few are given to help understand topics discussed later in this chapter. For details on XDC constraints and syntax, please refer to UG903 published by Xilinx.

14.1.1 Creating and Defining a Clock

create_clock Tcl command allows user to define clock on a certain port and also allows users to specify properties like period, waveform, root, etc. Unless a clock is defined using the *create_clock* command, static timing analysis is not performed on the clock. Also, *create_clock* command defines primary clocks, and all *derived* clocks are automatically inferred. Usually the *derived* clocks come from the clock modifying blocks like MMCM and PLL.

S. Dasasathyan (✉)
Xilinx Inc., San Jose, CA
e-mail: srini.das@gmail.com

S. Churiwala (ed.), *Designing with Xilinx® FPGAs*,
DOI 10.1007/978-3-319-42438-5_14

14.1.2 Defining Clock Relationships

Like all other SDC-based tools, Vivado also does timing analysis on all the cross-clock paths. However, designers in certain occasions would want to ignore certain paths, because those paths are either static paths (no signal transition happens) or the paths are asynchronous and hence should not be timed. In such cases *set_clock_groups* or *set_false_path* commands are used to preclude certain portions of the designs from timing analysis. This is an essential step as ISE (the previous Xilinx tool) which used UCF constraints, assumed the opposite, i.e., unless clock relationship was specified, timing analysis was not done on cross-clock paths.

14.1.3 Timing Analysis

Given these basic definitions of creating clock constraints and specifying clock relationships, Vivado's timing analysis engine does several checks under the static timing analysis engine. The timing analysis engine analyzes and reports slack at the timing path endpoints. The slack is the difference between the data required time and the data arrival time at the path endpoint. A data is safely transferred between two registers if both the setup and hold relationships are successfully verified on that path. In other words, if both setup and hold slacks are positive, the path is considered good from a timing point of view. The following are the checks performed by Vivado's timing analysis engine:

- Setup check
- Hold check
- Pulse-width check

14.2 Generating Timing Reports

The first step in timing closure is to understand whether the design has met all the timing checks or not. In order to generate timing reports to view failing paths, the following options are available in Vivado.

14.2.1 Report Timing Summary

Report timing summary gives an overall picture of timing on the design. It performs *setup*, *hold*, *pulse-width* checks, and gives a summary on whether some or all of these checks have failed. Even if one of the checks has failed, this command reports that the design has failed to meet timing. Based on this report, it can be decided if further steps are needed to achieve timing closure. Figure 14.1 gives a sample snapshot of the command, where *setup*, *hold*, and *pulse-width* violations are checked.

Design Timing Summary

Setup		Hold		Pulse Width	
Worst Negative Slack (WNS):	-1.120 ns	Worst Hold Slack (WHS):	0.016 ns	Worst Pulse Width Slack (WPWS):	0.381 ns
Total Negative Slack (TNS):	-6027.573 ns	Total Hold Slack (THS):	0.000 ns	Total Pulse Width Negative Slack (TPWS):	0.000 ns
Number of Failing Endpoints:	34121	Number of Failing Endpoints:	0	Number of Failing Endpoints:	0
Total Number of Endpoints:	490437	Total Number of Endpoints:	490437	Total Number of Endpoints:	154291

Fig. 14.1 Report timing summary output

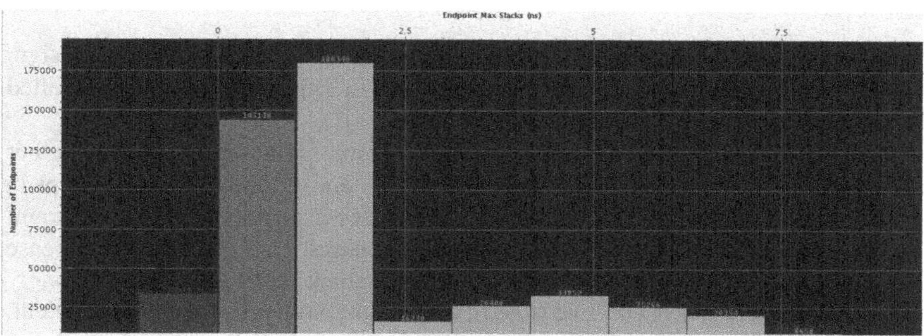

Fig. 14.2 Slack histogram

Once the design is determined to have not met timing requirements, you can further analyze failing timing paths in the design by running report timing or slack histogram command.

14.2.2 Report Timing

Report timing summary only gives a top-level report on timing failures; however, *report timing* gives details of all the paths that fail timing checks (setup and hold). By default *report timing* reports on all path groups and prints the top 10 paths in each path group and sort it by slack in ascending order. Additional filters can be added to customize timing analysis on different *from*, *through*, or *to* points as well as select more paths to view. *Report timing* only works for *setup* and *hold* checks. *Pulse-width* checks are reported in Vivado log file indicating where the errors are.

14.2.3 Slack Histogram

Another way to see the failing timing paths is to generate *slack histogram*. *Slack histogram* gives a concise view of all the timing paths across all path groups. Figure 14.2 shows a sample slack histogram plot. Slack histogram divides the slacks into different bins. The *X*-axis represents different slack bins and the *Y*-axis represents the number of paths in each bin. Clicking on each of the bars filters the paths in that bin, where you can examine paths in each of the bin.

In both report timing and slack histogram, you can click and double-click any of the paths to examine each of the timing path in detail, including characteristics of the path as well as placement and connectivity details.

14.3 Timing Paths and Constraint Correctness

Timing paths are defined by the connectivity between the instances of the design. In digital designs, timing paths are formed by a pair of sequential elements controlled by the same clock or by two different clocks.

In order to debug and fix the timing paths, it is important to first check whether these paths are valid or not. Checking constraints is one of the key and easy steps in getting to timing closure. One of the common issues in writing of XDC constraint is related to incorrect cross-clock domain crossing paths. Timer takes the worst case requirements for timing analysis. Hence if cross-clock paths are getting wrongly timed (very often they needn't be timed), they might have very tough requirement, resulting in a big negative slack. *Report CDC* and *report clock interaction* are two very useful commands to check if the interclock paths are being timed correctly.

14.3.1 Clock Interaction

Report clock interaction gives a matrix and specifies where all the clock pairs in the design are considered for interaction. Each entry in the matrix is color coded. All the entries across the diagonal are the paths within the same clock group. It is important to examine if there are any unexpected cross-clock domain paths, and fix them by adding proper XDC constraints (*set_false_path*, *set_clock_groups*). Xilinx published UG903 has more details.

14.3.2 Report Clock Domain Crossing

Report CDC (clock domain crossing) performs a structural analysis of the clock domain crossings in your design. You can use this information to identify potentially unsafe CDCs, which will lead to metastability or data coherency issues. While the CDC report is similar to the clock interaction report, the CDC report focuses on structures and their timing constraints, but does not provide information related to timing slack.

Before generating the CDC report, you must ensure that the design has been properly constrained and there are no missing clock definitions. *Report CDC* only analyzes and reports paths where both source and destination clocks have been defined. *Report CDC* performs structural analysis on:

1. On all paths between asynchronous clocks
2. Only on paths between synchronous clocks that have the timing exceptions (e.g., clocks coming out of MMCM)

Synchronous clock paths with no such timing exception are assumed to be safely timed and are not analyzed by the CDC engine. The report CDC operates without taking into consideration any net or cell delays.

14.4 Timing Closure Techniques

14.4.1 Critical Path Analysis

Timing reports can be generated at any stage during the synthesis and/or implementation phase. You should generate timing reports at each stage after synthesis, placement, and routing and analyze the paths to make sure that the design is converging. Catching and fixing issues earlier in the flow will save several iterations of the subsequent stages. For example, fixing issues at synthesis will save time in place and route stage.

A timing failure might happen due to multiple different reasons. Based on the analysis of the timing paths, fixes may be required at synthesis stage or the placement and routing stage. Hence it is important to study the characteristic of top failing paths to determine the reasons and fixes. Below are some of the important characteristics in the timing paths that can be examined and remedies that can be taken to mitigate them.

14.4.2 Logic vs. Wire Delay

Critical path delay can be broken down into logic delay and wire delay. The percentage of logic and wire delay in critical path can help to determine where to reduce delays. A low logic delay component usually means that wire delay is higher, where potentially floor planning the design can help in timing closure. A higher logic delay component means that there are too many logic levels in the design.

14.4.3 Reducing Logic Levels

For paths with higher levels of logic, looking at the levels of logic in the top failing paths can reveal if there are any issues in the RTL or inferring of the logic.

Synthesis step in Vivado infers structures in optimal way to balance between area and speed. Different RTL coding styles guide the tool to infer structures that are sometimes area optimal or performance optimal. By observing the logic levels in critical path, we can identify if we need to change either RTL coding style or guide the tool to infer for performance as opposed to area. To reduce the levels of logic, you can return to the RTL and check for the following general issues. In addition, refer to Chap. 9 for controlling synthesis behavior.

- Use *FSM_ENCODING* in your RTL to infer ONE_HOT FSM, which are usually better for speed.
- Use *CASE* statements instead of nested *IF-ELSE* statements; though the former takes more area, it has efficient inferences of Muxes which leads to better delays.
- Add pipeline registers to the critical path.

Any change to RTL will require resynthesizing the design. Several iterations may be needed to get optimal depth of logic.

14.4.4 Clock Skew

Clock skew is the difference between delays that clock takes from *common source* to capture flop/sequential element and the launch flop/sequential element. Examining the magnitude of clock skew can reveal issues in clocking structure. A design with high clock skew in critical paths usually means that the clocking structure needs to be revisited. Using MMCMs to multiply/divide clocks is recommended than using LUTs. UltraScale and newer devices have a very flexible clock architecture and offer lots of clocks to the user. To ease the issue of reducing clock skew and to generate *H-tree* clocking structures, the device offers *CLOCK_ROOT* which is the center tap points from where clock distribution happens. *CLOCK_ROOT* is chosen by Vivado for set of clock loads such that clock skew for the set of loads is minimal. However, in some cases where the paths are legal cross-clock domain paths, clock skew might be higher. In these cases user can choose *CLOCK_ROOT* manually to reduce the clock skew. UG912 from Xilinx explains the mechanism to modify *CLOCK_ROOT* location.

14.4.5 Reducing High-Fanout Signals

High-fanout signals typically pose a challenge to the place and route tools, as due to the very nature they have many connections, and the placement will be spread out. Due to this, delay on the net would be relatively higher. If the top several critical paths have some commonality that all of them involve high-fanout signal, some optimization can be done at RTL level to reduce the fanout coupled with options to synthesis tool. Some options are:

Duplicate the driver and tell the synthesis tool not to remove the duplicate logic (attribute *DONT_TOUCH*).

For the signals other than control signals such as reset, set, and clock enable, using *max_fanout* in synthesis will direct synthesis to replicate the driver.

Another option is to use *phys_opt_design* (post-placement). This command performs timing-based logic replication of high-fanout drivers and critical-path cells. Drivers are replicated, then loads are distributed among the replicated drivers, and the replicated drivers are automatically placed. This optional command can be run after placement and before routing.

14.4.6 Control Sets and Control Set Optimization

In Xilinx FPGA architecture (for 7 series and UltraScale), each *slice* has eight flip-flops (FFs). These eight FFs share control signals, so the FFs that are placed in the same slice should have same control sets. Hence the flops in the same slice have to share the control set. Placer algorithm honors this constraint by placing FFs of the same control sets together. Xilinx FPGAs can accommodate several thousand control sets; however, the higher the number of control sets, the more complex the job for placer to place flops into slices without wasting flops. *report_control_sets* command can be used to assess the number of unique control sets in the design. Under verbose options, the command gives details on the distribution of the fanouts of the control signal.

Vivado synthesis has an option which is used to specify threshold for synchronous control set optimization to lower number of control sets. The number set to this value specifies how large the fanout of a control set should be before it starts using it as a control set. For example, if *control_set_opt_threshold* is set to 5, a synchronous reset that only fans out to 5 registers would be moved to the *D* input's logic rather than using the reset line of a register. The default threshold value is currently set to 4.

Other ways to reduce control sets is to use *resets* judiciously. Be selective on the use of *resets* by observing the following points:

- Have resets only where they have impact on functionality.
- Use synchronous resets rather than asynchronous reset.

14.4.7 Floor Planning

Examining the critical path in the Vivado GUI will show the placement of the logic in the path. Sometimes, placer while trying to optimize several constraints might yield a suboptimal placement. Examining the top several critical paths in the GUI will give an idea if the placer indeed did a suboptimal job in placement of critical-path object. If so, floor planning can be done to guide the placer. A hierarchical floor plan can reduce the route delay in the critical logic. A good starting point when floor planning for the first time is to floor plan only the logic that the implementation tools consider timing critical. Generally start with the lower-level hierarchies that the place and route stage finds to be timing critical. More often it is useful to look at the placement of block RAMs and DSP blocks, as these are not distributed throughout the FPGA. Floor planning them not only gives better performance but also predictive results in future iterations of the same project. When the design meets timing, it is also possible to reuse the placement.

For SSI devices, floor planning poses additional requirements to consider, which are explained in Chap. 13.

14.4.8 Physical Optimization

Physical optimization performs optimization on the paths that fail to meet timing. Optimizations involve replication, retiming, hold fixing, and placement improvement. Physical optimization is usually run after placement when the timing picture is reasonably accurate. These optimizations are invoked by explicitly running the optional *phys_opt_design* command. This command performs the following physical optimizations.

High-Fanout Optimization: High-fanout nets, with negative slack within a percentage of the WNS, are considered for replication. The drivers are replicated and the replicated drivers are placed near to cluster of loads.

Placement-Based Optimization: Cells on the critical path are replaced to reduce wire delays.

Rewire: LUT connections are swapped to reduce the number of logic levels for critical signals. LUT equations are modified to maintain design functionality.

Critical-Cell Optimization: Cells in failing paths are replicated. If the loads on a specific cell are placed far apart, the cell may be replicated with new drivers placed closer to load clusters. High fanout is not a requirement for this optimization to occur, but the path must fail timing with slack within a percentage of the worst negative slack.

DSP Register Optimization: Registers are moved out of the DSP cell into the logic array or from logic to DSP cells if it improves the delay on the critical path.

Block RAM Register Optimization: Registers are moved out of the block RAM cell into the logic array or from logic to block RAM cells if it improves the delay on the critical path.

Retiming: Registers are moved across combinational logic to provide better timing.

Forced Net Replication: Net drivers are replicated, regardless of timing slack. Replication is based on load placements and requires manual analysis to determine if replication is sufficient. If further replication is required, nets can be replicated repeatedly by successive commands. Although timing is ignored, the net must be in a timing-constrained path to trigger the replication.

The above optimizations are run only during post-placement physical optimization steps; however, Vivado also allows to run physical optimization at post-route stage also. Only a subset of the optimizations are run at post-route stage, as the runtime of physical optimization post-routing is higher.

14.4.9 Strategy and Directives

Directives are powerful features that are available with every implementation step (synthesis, optimize design, placement, physical optimization, and routing). Directives give the implementation step to direct behavior of the algorithms toward alternate goal. It changes the implementation step by using:

- Different flows
- Different algorithms
- Different objectives

Directives allow each implementation step to enable more design space exploration than in the default mode. Directives have different objectives such as *reduce area, reduce runtime, improve performance, and improve power.*

Directives are enabled by running any synthesis and implementation step with the option *-directive*. Usually the names of the directive are chosen to indicate how different they are compared to the default behavior and their objective. Every implementation step has the directive *explore*. Explore allows the implementation step to work in a high effort mode to meet the timing objective at the expense of runtime. For designs with very tight requirements, it is recommended to use *explore* directive for most of the implementation steps (especially placement and physical optimization). Directives related to placement usually give the biggest improvement for performance. Please refer to UG904 from Xilinx for details on the list of directives and what each of the directive's objectives is.

Strategies define the flow of Vivado and customize the different implementaiton steps, and how each of these steps are configured. As each synthesis and implementation step has varieties of options and directives, strategies configure the best possible combination of these switches. You can also define your own custom strategy. Strategies are categorized into the following:

- Performance
- Area
- Power
- Flow
- Congestion

Each of the above strategy categories has several strategies which can be used to extract the last mile performance from the tools. In the context of timing closure, categories related to performance and congestion are applicable. One way is to run all the available performance strategies and pick the best results.

14.4.10 Congestion and Congestion Alleviation

FPGA routing architecture has different kinds of routing resources to service different scenarios seen in placement of the design. Congestion can happen when in a region there is more demand of certain kinds or all kinds of routing resources than their availability. Extent of the congestion regions defines whether the congestion is local or global. Router and placement algorithms, in order to alleviate congestion, introduce *white spaces* and *detours*. These changes may impact the routing delays by worsening them, which impact the timing of the design. There are certain steps you can take to reduce the effect of congestion on timing. Congested regions can be determined by running congestion reporting using *report design analysis*. Also designs with heavy utilization of block RAMs, MuxF7s, and MuxF8s and distributed RAMs have a tendency to have congestion. Care should be taken to reduce the utilization of any block with high connectivity. Blocks with high connectivity increase number of signals coming in a region where the blocks are placed. If there are many high connectivity blocks placed in a small region, one can increase the size of a region by defining a *pblock*. The size of the pblock can be increased to make it large enough to have enough routing resources to complete routing all nets and thereby alleviating congestion.

14.4.11 Report Design Analysis

Report design analysis is a command that summarizes several important details on the critical paths. Commonly occurring issues in critical paths are summarized in a tabular format. By looking at the characteristics of several critical paths, issues can be deduced. *Report design analysis* has three modes of operation:

- Timing
- Congestion
- Complexity

Timing mode is used to find out the characteristics of critical paths. For each of the path, many important characterisitcs are printed. For example, it is easy to determine if the top critical paths have block RAMs and whether they are registered or not. Or, if the top several critical paths have LUTs which are combined in synthesis stage (we can turn this off by using *-lc off* option). Xilinx published UG906 provides information on other meaningful information that can be obtained from this report.

Congestion mode gives the post-placement and post-routing congestion windows, and *complexity* computes the *rent's* exponent of the netlist or modules specified. Congestion combined with complexity can determine whether the netlist itself is inherently congested, or the congestion is placement induced. Using congestion mode, you can find the congested window and also determine what modules are placed in the region. Later you can run complexity on these modules and compute the *rent's* complexity on them. Rule of thumb says that any rent's complexity over *0.7* can be considered as an issue in netlist.

14.4.12 Timing Closure and Hold Violation

The previous section covered several techniques related to closure of timing which mainly focused on setup violations. Hold violations are also another kind of timing failures that you need to be aware of. Hold violations are severe, as reducing the clock frequency will not help in timing closure. Vivado tool is hold aware and tries to mitigate the violations by detouring and adding extra delay to the paths failing *hold.* However, you should be aware of these requirements and not solely depend on tool to fix the issues. Buffers can be added in hold failing path with *DONT_TOUCH* attribute so that synthesis tool does not optimize them away. Further post-route physical optimization and few router directives can also help to reduce the hold violation. Figure 14.3 provides a top-level flow chart for achieving timing closure on your design.

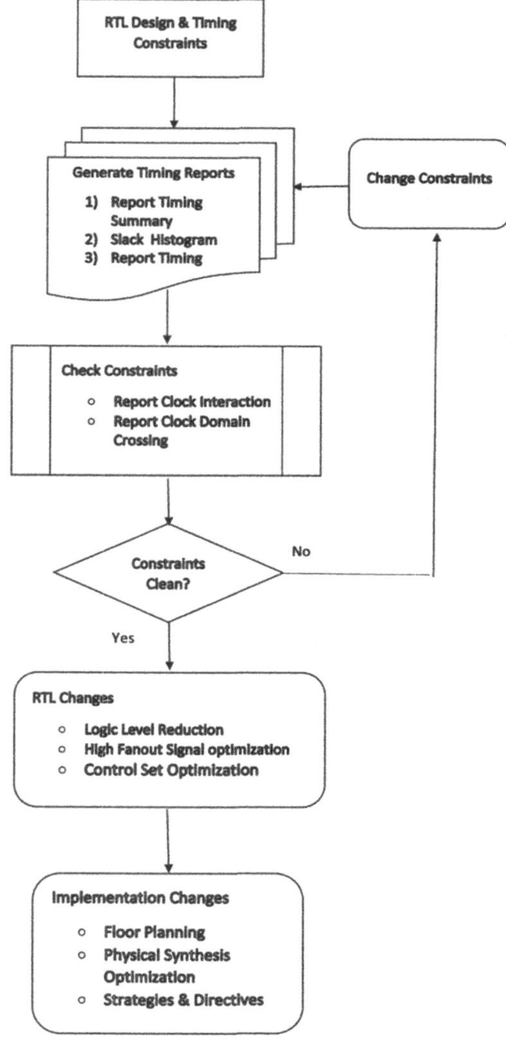

Fig. 14.3 Flow chart for timing closure

Chapter 15
Power Analysis and Optimization

Anup Kumar Sultania, Chun Zhang, Darshak Kumarpal Gandhi, and Fan Zhang

15.1 Introduction

There are several factors that influence the power consumption of a given system and can be divided into two broad categories—physical and functional. Board design, packaging, and device selections are examples of physical factors, whereas functionality is largely related to the RTL design itself. In this chapter, we will explore the tools available for power estimation and optimization.

Power estimation can be done at various levels of granularity, and the accuracy of the estimation is dependent on the amount of information you can provide. The more information you can provide, the more accurate the estimates will be compared to the power consumption on the final hardware. Xilinx provides three tools to help analyze and optimize for power (see Fig. 15.1). These are:

1. Xilinx Power Estimator (*XPE*): This is used for predesign phase estimation. This is an Excel-based tool and relies heavily on user-entered information in both physical and functional categories. While *XPE* is very helpful in doing power budgeting in the early phase of a project, it can also be used to do a what-if analysis for an implemented design.
2. Vivado *Report Power*: This is used for post-design phase power analysis. This is a more accurate tool as it operates on a synthesized, placed, or routed netlist. While majority of the functional information is obtained from the netlist, you still need to enter the physical factors and switching activity information to get an accurate power estimation.

A.K. Sultania (✉) • C. Zhang • D.K. Gandhi • F. Zhang
Xilinx Inc., San Jose, CA, USA
e-mail: anup.sultania@gmail.com

S. Churiwala (ed.), *Designing with Xilinx® FPGAs*,
DOI 10.1007/978-3-319-42438-5_15

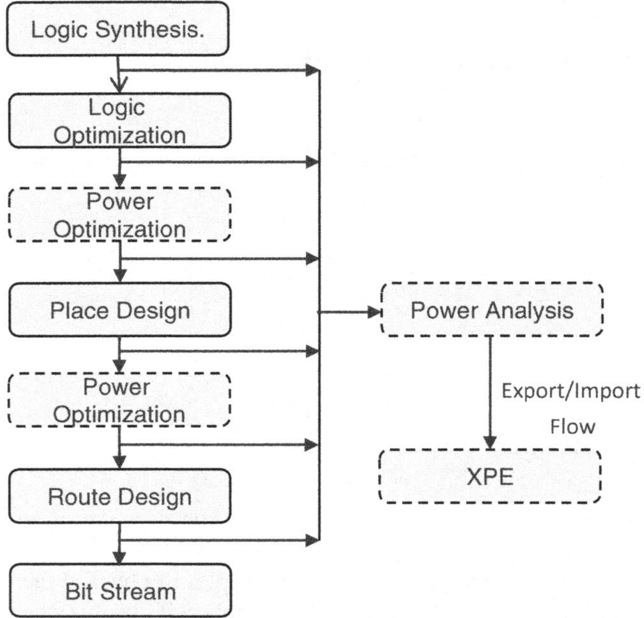

Fig. 15.1 Xilinx power analysis and optimization flow

3. Vivado Power Optimization: This implements ASIC style clock-gating technique based on sequential analysis of the designs. It reduces the activity on portions of the design that do not impact the design output.

FPGA power can vary from few hundreds of mW to tens of Watts. It depends on a variety of factors – design function, clock frequency, switching activity, and board and environmental setup. Power on an FPGA can be broadly divided into four categories:

- Device Static: This is the power which is consumed even if there is no design configured into FPGA. This is typically measured by programming a blank bitstream into the device and is a function of process, voltage, and temperature.
- Core Dynamic: This is the dynamic power consumed when the FPGA is in use and does not include I/O and Transceiver power.
- I/O and Transceiver: Power in I/O and Transceivers is categorized separately as they have a high impact on overall power. The tools provide a capability to explore various configurations to make the best possible decision from power standpoint.

15.2 Xilinx Power Estimator (XPE)

XPE is a predesign phase tool meant to be used early in the project cycle to come up with power budgets for FPGA. It also helps Xilinx to provide customers an opportunity to explore power profiles for future devices. In the backend, XPE implements power models which take in user-entered information and generate a power number. The power models go through multiple stages—*Preview* for models based on early device design specification, *Advanced* for models based on device design simulation, *Preliminary* for models based on measurements on early silicon, and *Production* for models based on production silicon measurements.

XPE being Microsoft Excel-based tool, it retains a majority of Excel capabilities. It is divided into several sheets; several of them are dedicated to a specific resource type on the FPGA. On *Summary* sheet, device selection and environment setup can be explored. It also contains detailed power report. *Snapshot* sheet allows to compare power reports between different settings. There is also a blank *user* sheet which retains all the Excel functionality. It can be used in a variety of ways, from scratch space to detailed system level block diagram, and can cross-reference data from rest of the sheets.

Apart from exploring different device and thermal setup for optimal static power, it is also important to explore the relative dynamic power impacts across different configurations of various blocks. For example, *Transceiver* sheet mainly asks for basic transceiver-related inputs like *channel count*, *data rate*, *data width*, and *operation modes*. Besides these, you can do a what-if analysis to see power savings of choosing low-power mode (*LPM*) over decision–feedback equalization (*DFE*). This sheet can also estimate the additional power of using eye scanning, out of band (OOB) sequence generation, or any hard IP blocks with a given transceiver.

One more example is the *I/O* sheet which asks for basic I/O characteristics like *data rates*, *toggle rates*, *enable rates*, and *pin configurations*. It also gives the capability to do a what-if analysis between high-performance (*HP*) and high-range (*HR*) I/O banks. It gives an extensive and intelligent drop-down list of IO standards depending on availability in selected I/O bank and device. For more accurate estimation, advanced users can also provide input termination and output impedance when they are supported by selected I/O standard.

Manually entering entire design data in XPE can be tedious and confusing at times. To aid in this, XPE provides various wizards—*Quick Estimate*, *Memory Interface Configuration*, *Memory Generator*, and *Transceiver Wizard*. Quick Estimate wizard is to do a very quick and coarse power estimation. The remaining wizards are for ease of design data entry. For example, you can use *Transceiver* Wizard to choose from a variety of protocols from the drop-down menu and enter few key information like *data rate*, *clock*, etc., and it will not only populate the *Transceiver* sheet but also add link layer logic information in the *Logic* sheet. XPE also allows you to delete the design data added through one of the above wizards by using *Manage IP* wizard.

As a final note, XPE can only be as accurate as the data entered. Often, it is very difficult to estimate power accurately because accurate switching activity and design information is not known very early in the design cycle. If sufficient information is provided, XPE can estimate *device static*, *I/O*, and *Transceiver* power with reasonable accuracy. However, it still does not have sufficient design connectivity information to accurately estimate *core dynamic* power. Since power budgets are frozen early in the design cycle, it is important to account for this uncertainty early on.

15.3 Vivado Report Power

Report Power is a very detailed power analysis tool and computes power at a fine-grained level. For example, it estimates power for each *LUT* based on switching activity and capacitance information present at input and output pins of the *LUT*. Similarly, it accounts for the exact routing of each net while estimating power. This is in contrast to the coarse model present in XPE.

The Vivado Power Analysis engine uses four types of information as shown in Fig. 15.2. It gathers the netlist information and configurations of various blocks by analyzing the design. It uses hardware characterization data based on selected device and package. Operating conditions like process, voltage, and temperature must be set and are typically pre-decided when exploring through XPE. Finally, power constraints comprising of switching activity constraints and clock constraints need to be carefully set to get accurate power estimation. All these are passed as inputs to various algorithms to come up with a detailed power estimation.

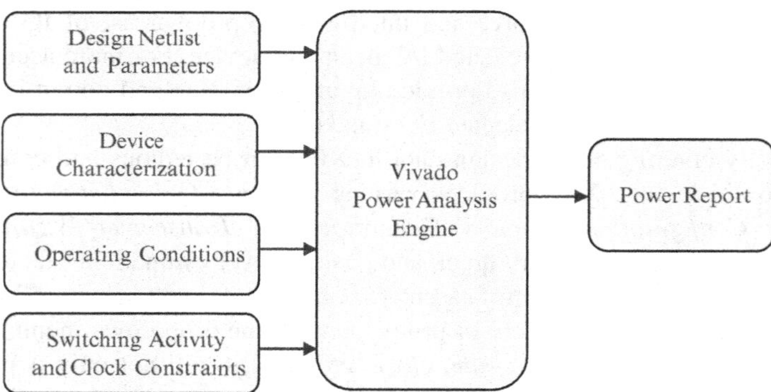

Fig. 15.2 Vivado power analysis engine

15.3.1 Operating Conditions

Defining proper operating conditions are essential for the accuracy of power calculations. Power engine can use predefined typical or calculated values for most of the operating conditions; however, it is strongly recommended that you overwrite a few critical values based on the system specifications. For example, if you are aware of the maximum junction temperature, then you should set that in operating conditions. This will prevent the tool from estimating junction temperature based on environment and board setup. Similarly, the tool default for the process corner is *typical.* You should change this to *maximum* to get worst-case device static power. You should also provide exact or worst-case (i.e., maximum) supply voltage values provided by external power regulators as power depends significantly on voltage supply values.

15.3.2 Power Constraints

Similar to static timing analysis (STA) tools, Vivado Power Analysis (report power) requires you to provide power constraints to guide the tool for accurate power prediction. Power constraints are specific to clock frequency and switching activities. For clocks, the frequency can be constrained using the same SDC timing commands. You need to guarantee that all the clocks are properly constrained. Switching activity is represented by a pair of values as (*toggle_rate, static_probability*). By definition, *toggle_rate* is the probability of a signal in a synchronous design making a '0'→'1' or '1'→'0' transition within a clock cycle. *Static_probability* is the probability of a signal being *1* in any clock cycle. Figure 15.3 shows signal *x* with *toggle_rate* of 40 % and *static_probability* of 0.3 within a ten clock-cycle window.

Power analysis requires switching activities for all nets. At the first appearance, this seems like a daunting task for users to provide all switching activity constraints in brute force. The novel methodology in Vivado Power Analysis requires you to only provide switching activities for a subset of nets rather than all of them and, together with the activity propagation engine (see Sect. 15.3.3), greatly minimizes design effort and at the same time provides accurate results. There are two ways to provide switching activity information.

First, you can simulate the design (or its portions) to generate switching activity constraints. It is recommended that you do simulation on some critical modules of

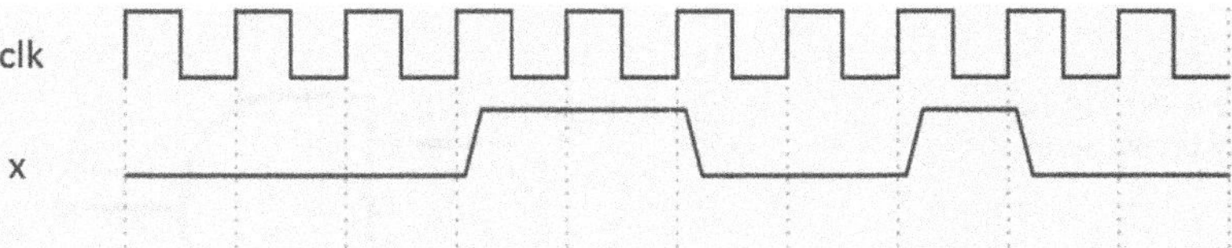

Fig. 15.3 Signal toggles four times within ten clock cycles and stays "1" for 3 out of 10 cycles

the design and generate a Switching Activity Interchange Format (*SAIF*) file. This file can then be used to annotate switching activities on the design. Power results are greatly impacted by simulation done at different design stages as well as with or without glitches. For accuracy purpose, simulation at post route stage with delays will generate switching activities most close to real hardware.

Second, if simulation results are not available, you can constrain critical control signals of the design and let activity propagation engine estimate activities on the remaining nets. The critical control signals are those that can enable or disable a large portion of the design. Examples of critical control signals include *set/reset* pins that drive large flop fan-outs, block RAM *enable* pins that switch on/off the data path, clock selection pins to switch between clocks at clock controller output, pins that enables the power down or sleep feature of hard IP blocks, etc. Not all control signals are critical. Control signals that only reset limited number of nonessential flops can be safely ignored without impacting the accuracy of power prediction.

In addition to critical control signals, another way of guiding the tool is to provide activities on groups of data path signals, for example, block RAM or GT data output pins and chip-level input ports. This approach of setting activities en masse is useful in doing worst-case power estimation.

15.3.3 Activity Propagation

After constraints are provided to annotate partial design nets with switching activities, activity propagation engine triggers to propagate activities on the remaining parts of the design. The activity propagation is a statistical analysis based engine and, on a large design with a million LUTs and registers, can usually complete within several minutes. Figure 15.4 demonstrates activity propagation for a simple AND gate.

Assume the same *static_probability* for input *a* and input *b*: $SP(a) = SP(b) = 0.5$. Those values can be from user constraints or propagated values from previous logic. In the case of the *AND* gate, $SP(o)$ is computed to be $0.5 \times 0.5 = 0.25$. This is under the assumption that inputs *a* and *b* are totally independent of each other. Similarly, activity propagation engine will also compute output toggle rate. Details of this algorithm are not necessary for you as a user of these tools. Not just combinational circuits, activity propagation engine can also propagate activities across sequential circuits.

Real designs are usually large and complex due to correlations between different signals. It is infeasible to compute exact switching activity for all the nets within a

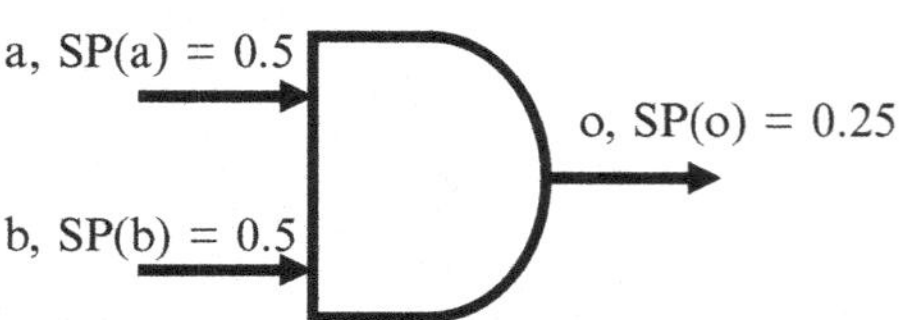

Fig. 15.4 AND gate activity propagation

reasonable amount of time. *Report Power* activity propagation engine is able to intelligently solve a subset of correlations in the design and trades off between runtime and accuracy. With proper constraints on clocks and critical control signals, *Report Power* is able to predict power reasonably close to hardware measurement. It is to be noted that activity propagation engine does not override user-provided constraints, rather it uses them as inputs to estimate activities on remaining nets.

Because of its ease of use over other activity analysis methods like simulation, activity propagation can be used to efficiently evaluate relative impact on power after a design netlist change or switching activity change.

15.3.4 Export–Import Flow with XPE

Vivado *Report Power* and *XPE* are two independent power analysis tools. In *Report Power*, majority of the information is gathered directly from the design where as in *XPE* you have to enter all the information. When running *Report Power*, you can export all the physical and functional information to an *XPE* exchange (*.xpe*) file which can be easily imported into *XPE* tool. While *XPE* has very high-level design information and less model accuracy compared to Vivado, exporting design information from *Report Power* to *XPE* can be very helpful for multiple use cases.

One use case for this flow is to do a what-if analysis at post synthesis design stage. When the power reported in *Report Power* exceeds allocated budget, the design can be exported to XPE to evaluate power saving ideas without actually making any RTL changes. For example, you can evaluate how much power reduction can be achieved by reducing the resource usage or changing configurations of blocks like BRAMs and DSPs. The impact of using different parts or environment settings and different voltage options can also be studied very easily. Snapshot and graph features of XPE come in very handy while doing several what-if analysis.

Another use case is to do a more accurate early estimation for the next generation of the design, which may reuse some design components from the current design. For example, if the next-generation design is going to use most of the similar design elements, then you can import the current *.xpe* file to XPE and make changes to sheets where the design change is predicted. Often the next generation of Xilinx FPGAs are supported in XPE relatively earlier than Vivado. In such cases, export–import flow is very helpful to study power profiles on not only existing devices but also future devices.

15.4 Vivado Power Optimization

Vivado power optimization exploits a variety of techniques to reduce the dynamic power consumption of the design. As shown in Fig. 15.5, it detects the clock cycles under which certain sequential circuit elements do not contribute to observable

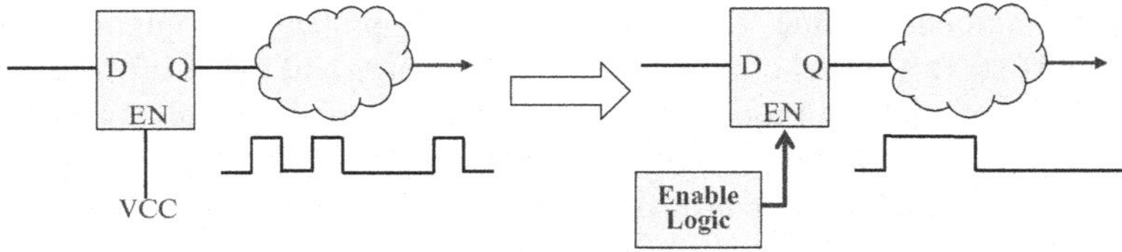

Fig. 15.5 Vivado power optimization

design functionality, and applies ASIC-like clock-gating techniques to reduce their activities. Due to the fact that FPGAs have dedicated clock routing resources, the clock gating is actually applied to the *enable* port of sequential elements such as a flop or block RAM. Compared to the coarse-grained clock gating that requires a non-trivial amount of design effort, Vivado power optimization is capable of automatically inferring more fine-grained gating conditions across multiple levels of logic and sequential boundaries.

15.4.1 Optimization Paradigms

The fundamental of Vivado power optimization is the inference of logic conditions under which the sequential element can be disabled without disturbing observable design states and/or functionalities. There are two major paradigms that Vivado power optimization explores: the output don't care (*ODC*) paradigm and the input don't toggle (*IDT*) paradigm. A brief introduction of these paradigms will help in intuitively understanding the potential netlist-level changes applied by Vivado power optimization, which is important for designing and analyzing low-power systems.

The *ODC* paradigm infers the enable condition by exploring the output side of a sequential element, with the key idea that the sequential element only needs to be enabled when its output is consumed by logic in the fan-out cone. As shown in Fig. 15.6, the output of *FF1* becomes don't care when *FF2*'s *CLR* signal is asserted. Consequently, Vivado power optimization infers that *FF1* only needs to be enabled when *FF2*'s *CLR* signal is de-asserted and applies that signal to the *enable* port of *FF1* through the inverter. Since a flop's *enable* decides its output data availability in the next clock cycle, the actual *enable* of the *FF1* needs to be traced back by one clock cycle which is applied through *FF3* in the example.

To infer *enable* conditions across sequential boundaries, Vivado power optimization performs multiple iterations of ODC analysis. This essentially unrolls the time span and back propagates ODC conditions across multiple levels of flops. In the example shown in Fig. 15.7, the ODC *enable* for flop *FF2* is inferred in the first iteration from its output observability at the MUX, while the ODC enable for *FF1* is decided in the second iteration based on *FF2*'s inferred ODC enable.

On the other hand, the IDT paradigm searches *enable* condition by exploring the input side of a sequential element, with the idea that if its input data remains same,

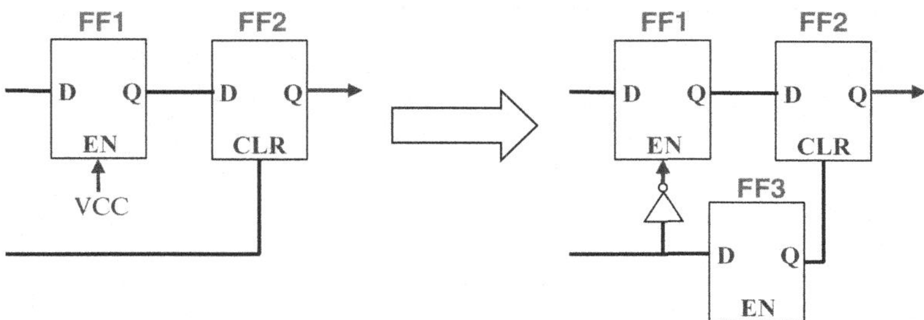

Fig. 15.6 ODC optimization paradigm

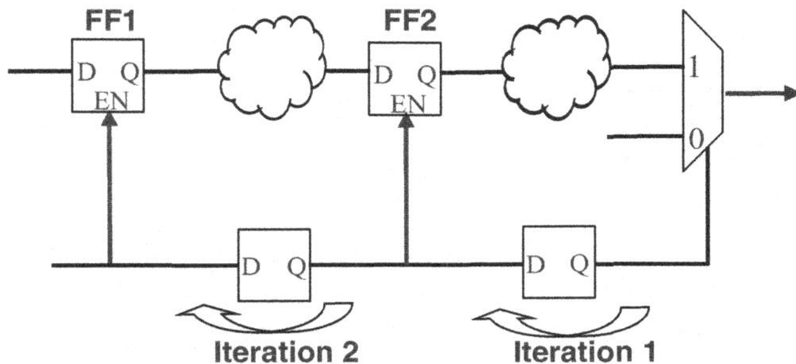

Fig. 15.7 Multiple iterations of ODC

the sequential element can be safely disabled without altering the direct output. In the example illustrated in Fig. 15.8, the flop *FF1*'s input doesn't toggle when $a=1$ and $b=0$ since its output is directly fed into its input. Consequently, Vivado power optimization generates the IDT enable signal of *FF1* as the complement of such disable condition, i.e., $EN = {\sim}a + b$. Generally speaking, the IDT paradigm is useful for reducing dynamic power of designs with many feedback loops.

In addition to the general ODC and IDT paradigms, Vivado power optimization also takes care of applying specific optimization techniques to certain high-power-consuming components such as block RAMs. To illustrate a few, the following techniques are deployed:

- Block RAM Structural ODC Optimization—Different from the general ODC paradigm, this optimization searches the conditions under which the block RAM is used in *write-only* manner and thus directly utilizes the *write-enable* signal as the block RAM's global enable control to suppress any unnecessary READ operations.
- Block RAM Write-Mode Optimization—Write mode defines the behavior of the block RAM's outputs when data is being written into it, which can be set to *NO_CHANGE* to suppress any unnecessary output toggling. To fully utilize this feature, Vivado power optimization searches the block RAMs whose outputs aren't consumed during WRITE operations and sets their write mode to *NO_CHANGE*.

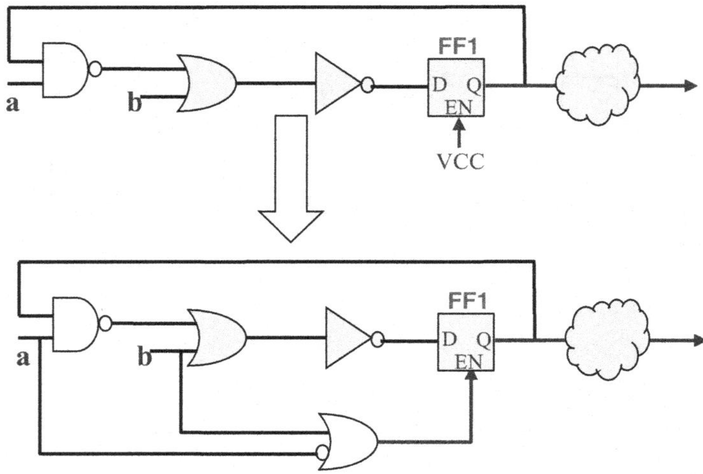

Fig. 15.8 IDT optimization paradigm

- Block RAM Quiescent IDT (QIDT) Optimization—When the block RAM's input addresses remain the same in two consecutive READ cycles, Vivado power optimization safely disables the block RAM without disturbing its functionality.
- Cascaded Block RAM Optimization—When multiple block RAMs are cascaded, only one of the block RAM needs to be active at the same time. Consequently, Vivado power optimization generates the enable signal for each block RAM in the cascaded chain from most-significant bits (MSBs) of the address bus such that only the block RAM being accessed is enabled.

Post Vivado power optimization, you may observe different outputs of certain sequential elements such as flops or block RAMs from simulation. This is expected since the activities of these elements are reduced by clock gating the *EN* port. But, Vivado power optimization guarantees that the design's observable functionality remains undisturbed (i.e., from primary outputs), since these sequential elements are only disabled during the clock cycles when their outputs are not consumed or remain unchanged.

15.4.2 Suggestions for Low-Power Design

In addition to automatically reducing the power consumption of the design, some design level considerations could further improve the power characteristics and/or create more optimization opportunities for Vivado power optimizer. In this subsection a few techniques good for low-power design are proposed:

- Cascaded block RAMs—To implement the same memory, block RAMs can be cascaded in different ways which impact the power and timing of the design. For example, to implement the 36*8K memory, one option is to have nine 4*8K block RAMs in parallel, each contributing 4 bits of the data. Although this achieves the highest speed, it requires all block RAMs to be active concurrently, which consumes a significant amount of power. On the other hand, the same memory can be implemented by cascading nine 36*1K block RAMs, which is power optimal since only one block RAM is active at the same time. Generally, you shall consider the balance between parallel/cascaded block RAM implementation to achieve the best power and speed trade-off.
- Distributed RAM vs. block RAM—Similarly, the choice of using distributed RAM or block RAM to implement the memory could affect the power consumption of a design. For instance, to implement the 32*100 memory, using one block RAM is functionally correct but wastes a large portion of the data capacity of the block RAM. On the other hand, the same memory can be implemented by 100 distributed RAMs without wasting any resource or power. Consequently, it is also a good idea to consider distributed RAM vs. block RAM under certain power and resource constraints.
- MUX chain design—The structure of the MUX chain decides the way ODC analysis is being performed by Vivado power optimization. Pushing the high-power-consuming element such as block RAM to the end of the MUX chain increases the chance for Vivado power optimization to find the best ODC enable condition for that element.
- XOR tree design.—While XOR tree is good for implementing arithmetic logics, it has the disadvantage of causing excessive glitches with increased levels of XOR gates. Consequently for power-centric designs, it is suggested to limit the logic levels of XOR tree by applying techniques such as inserting pipeline stages in between.

Chapter 16
System Monitor

Sanjay Kulkarni

16.1 Usage and Need

Since the introduction of Virtex5 FPGA devices, the *SYSMON* (*System Monitor*) has been a part of every new FPGA family introduced by Xilinx. The *SYSMON* allows you to monitor the critical parameters like on-chip temperature, voltages, power, etc. With each new generation of FPGA families (6-series, 7-Series, UltraScale, Zynq, Zynq US+ MPSoC, etc.), Xilinx has improved the capabilities of the *SYSMON* to cater to the newer challenges and user design requirements. You may expect similar trend for SYSMON features in Xilinx future FPGA families.

This chapter provides the details of System Monitor based upon the Xilinx latest UltraScale FPGA family.

16.2 Overview of SYSMON

The SYSMON functionality is built around the hard silicon block *ADC* (analog-to-digital converter) and its interface to various on-chip sensors. When combined with a number of on-chip sensors, the *ADC* is used to measure FPGA's physical operating parameters like on-chip power supply voltages and on-die temperature. The ADC provides a general-purpose, high-precision analog interface for a range of applications. The external analog interface inputs allow the ADC to monitor the physical environment of the board or enclosure. As soon as the FPGA is powered up, even before it is configured for any application purpose, the SYSMON is already activated and starts functioning. At this point, its functionality is restricted to the

S. Kulkarni (✉)
Technical Manager - Applications, Microsemi India Pvt. Ltd., Serilingampally Mandal, Kapil Towers, 13th Floor, Survey No. 115/1, Nanakramguda, Hyderabad, Telangana, 500032, India
e-mail: sanjukulkarni12@gmail.com

S. Churiwala (ed.), *Designing with Xilinx® FPGAs*,
DOI 10.1007/978-3-319-42438-5_16

measurement of on-chip parameters only. This data can be accessed through JTAG or with dedicated I2C interface. Even if the SYSMON is not part of the FPGA-based design, it is still accessible through these interfaces.

To understand further about SYSMON, let us look at the block diagram as shown in Fig. 16.1.

Access to external analog world is provided through a dedicated analog input pair (*VP*/*VN*) and 16 user-selectable analog inputs, known as *Auxiliary Analog* inputs. The ADC supports differential sampling of unipolar and bipolar analog input signals. The ADC has different range of operating modes to handle the external analog inputs. SYSMON block includes a rich set of *Configuration Registers*. These registers are classified into different groups like *Control Register*, *Alarm Register*, and *Status Register*.

SYSMON operates at very low-voltage level in UltraScale FPGA devices either using the external reference source (1.25 V) or on-chip voltage source (VCCAUX 1.8 V). If the requirement is restricted only to monitor the on-chip temperature and voltages, then it is always beneficial to use the on-chip voltage source as reference.

16.3 Evolution of SYSMON in Xilinx FPGA

Table 16.1 shows the comparison of SYSMON evolution in Xilinx 7 series through UltraScale+ FPGA families.

16.4 Using the SYSMON in System Design

As per the requirements for the system design, you can directly instantiate the hard macro template in the design RTL files and use appropriate interface ports as per the required configuration. The RTL instantiation template is available in Vivado under *Advanced Device Primitives*. Xilinx also provides the *SYSMON Wizard* IP core under Vivado *IP catalogue*, which helps the user to pre-configure the SYSMON for FPGA-based system applications. Figure 16.2 represents the UltraScale-based SYSMON primitive level diagram showing ports of different groups and interfaces.

It is expected that you should be familiar with each of the SYSMON ports group and its intended usage in their application when SYSMON is incorporated in the design. The *Dynamic Reconfigurable Port* group is mainly used to access the internal set of registers. *Control and Reset* port group is used to control the reset as well as controlled conversion start access. The *External Analog Input* ports are the main ports for you to connect up to 16 external analog channels to be monitored, while the *VP*/*VN* are dedicated input ports. The 16 analog inputs are actually FPGA general-purpose IOs, while the *VP*/*VN* pins are non-shared pins. If *VP*/*VN* is not used, then these pins cannot be used for any other general-purpose IO usage and

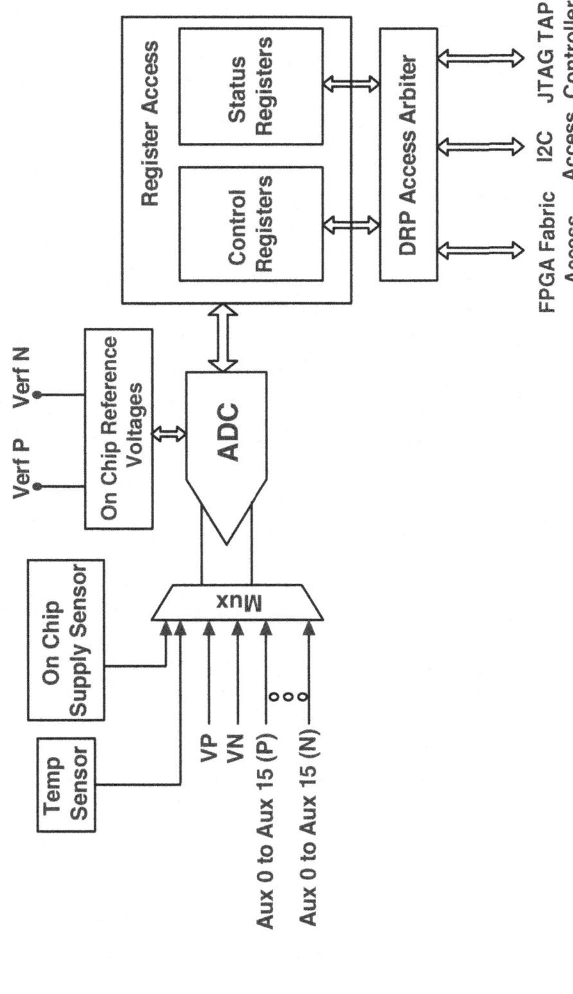

Fig. 16.1 Simplified block diagram of UltraScale FPGA-based system monitor

Table 16.1 Evolution of SYSMON in Xilinx family of FPGA devices

Feature	7-Series/Zynq FPGA (28 nm)	UltraScale FPGA (20 nm)	UltraScale+ FPGA (16 nm)
SYSMON primitive name	XADC	SYSMONE1	SYSMONE4
Resolution	12-bit	10-bit	10-bit
Sample rate	1 MSPS	200 KSPS	200 KSPS—In PL[a]
			1 MSPS—In PS[b]
Analog-to-digital converters	2 (in PL)	1 (in PL)	2 (1 in PL[a], 1 in PS[b])
Banks supporting external analog inputs	1	All IO banks support analog inputs	All banks (In PL[a] SYSMON only)
Alarm outputs	Total 8	Total 16	Total 16
	ALM[7:0]	ALM[15:0]	In PL including Supplies + Temp + Analog Bus
Supply sensors	System supply sensors:	System supply sensors:	System supply sensors:
	V_{CCINT}, V_{CCAUX}, V_{CCBRAM}	V_{CCINT}, V_{CCAUX}, V_{CCBRAM}	PL: V_{CCINT}, V_{CCAUX}, V_{CCBRAM}
	Zynq: V_{CCPINT}, V_{CCPAUX}, VC_{CO_DDR}		Zynq: $V_{CC\_PSINTLP}$, $V_{CC\_PSINTFP}$, V_{CC_PSAUX}
	User supply sensors:	User supply sensors:	User supply sensors:
	0	4	4 (Vuser[3:0])
			PS: many supplies (do not have the list)
Reconfiguration interfaces	Fabric DRP, JTAG TAP	Fabric DRP, JTAG TAP, I2C DRP	Fabric DRP or dedicated PS DRP, JTAG TAP, PMBUS, I2C DRP
Sequence mode	Default, single pass, continuous, single channel, simultaneous sampling, independent ADC	Default, single pass, continuous, and single channel	Default, single pass, continuous, dual sequence

[a]PL = programmable logic
[b]PS = processing system

need to be connected to analog ground. SYSMON also provides indicators for any adverse conditions through *ALARMS* group of signal. Your application can connect this group of signals to the monitoring LEDs on the board. The *Status Group* of signals mainly indicates the end of current conversion cycle, which channel is at present under process, status of JTAG access of registers suit, etc. The *I2C DRP* group of signals is used to provide two-wire standard low-cost *I2C* protocol access by external *I2C* master. These are the main peripheral ports you need to be aware while planning the system-level designs along with SYSMON.

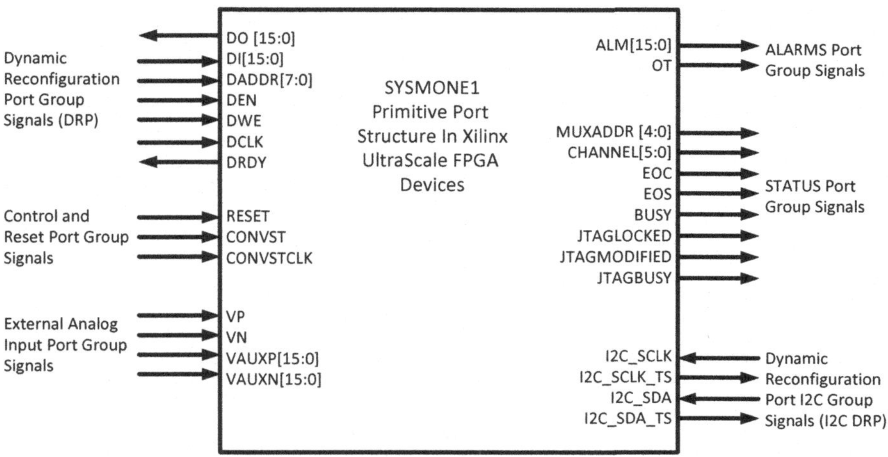

Fig. 16.2 SYSMONE1 primitive port structure in the UltraScale FPGA devices

16.5 ADC Capabilities of SYSMON

The heart of UltraScale FPGA-based SYSMON is designed around a 10-bit *ADC*, capable of working at 200 KSPS (kilo samples per second). It is commonly used between the external analog signals as well as on-chip sensors. The SYSMON ADC has nominal external input voltage range of 0–1 V. Various operating modes of the ADC, sensors, and analog inputs can be configured using the SYSMON *Control Registers*. The ADC supports *Unipolar Mode* (which is default mode) of operation for on-chip sensors, while for external channels both the *Unipolar* and *Bipolar* operation modes are supported. The ADC always produces 16-bit conversion result, out of which the 10 MSB (left most bits) represent the 10-bit transfer function, which is stored in the *Status Registers*. The remaining six LSB can be used to improve the resolution through averaging or filtering.

In case of *Unipolar Mode*, for input of 0 V, the ADC produces *0x000h* code, while for the highest input of 1 V the ADC produces full-scale code of *0x3FFh*. This shows that the ADC output in *Unipolar Mode* is straight binary equivalent. Each bit increase represents 977 μV increase. When external analog inputs are configured as *Bipolar Mode*, they can accommodate true differential and bipolar analog signals. The output coding of ADC in *Bipolar Mode* is two's complement. In this case also each of the bit count represents 977 μV. The diagrammatic representation of the *Unipolar* and *Bipolar Modes* transfer function of SYSMON is as shown in Fig. 16.3 and 16.4. Note the upper range of binary code representation is the same for *Unipolar* and *Bipolar Mode*. The output coding of the ADC in *Bipolar Mode* is two's complement and indicates the sign of the input signal on VP relative to VN.

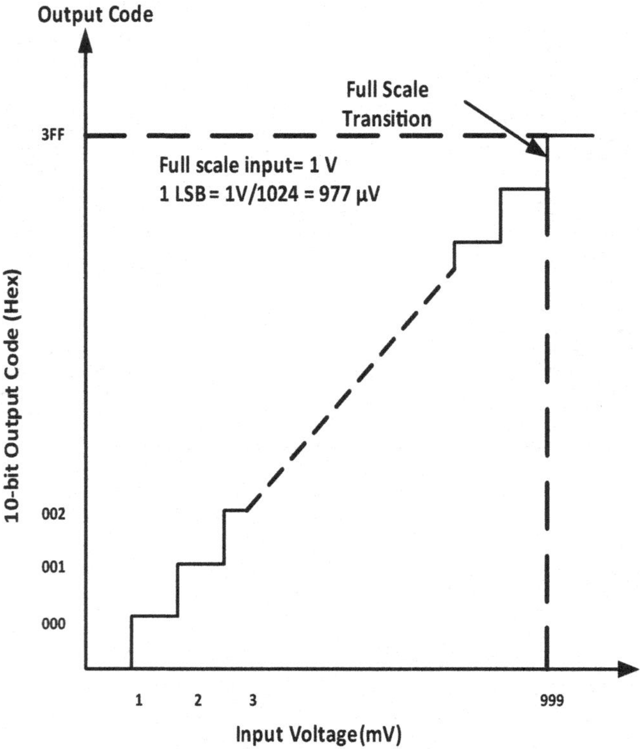

Fig. 16.3 Unipolar transfer function of SYSMON ADC

16.6 Transfer Function of Various On-Chip Sensors

16.6.1 Temperature Sensor

The UltraScale based SYSMON has on-chip temperature sensor, which produces the equivalent output voltage proportionate to the die temperature. The ADC can be configured to use external as well as internal reference voltage for temperature conversion. When using an external reference voltage, the transfer function of the temperature sensor is as given by (16.1). ADC value corresponding to a given temperature can also be obtained by the same equation:

$$T = \frac{\text{ADC} \times 502.9098}{2^{\text{bits}}} - 273.8195 \tag{16.1}$$

where:

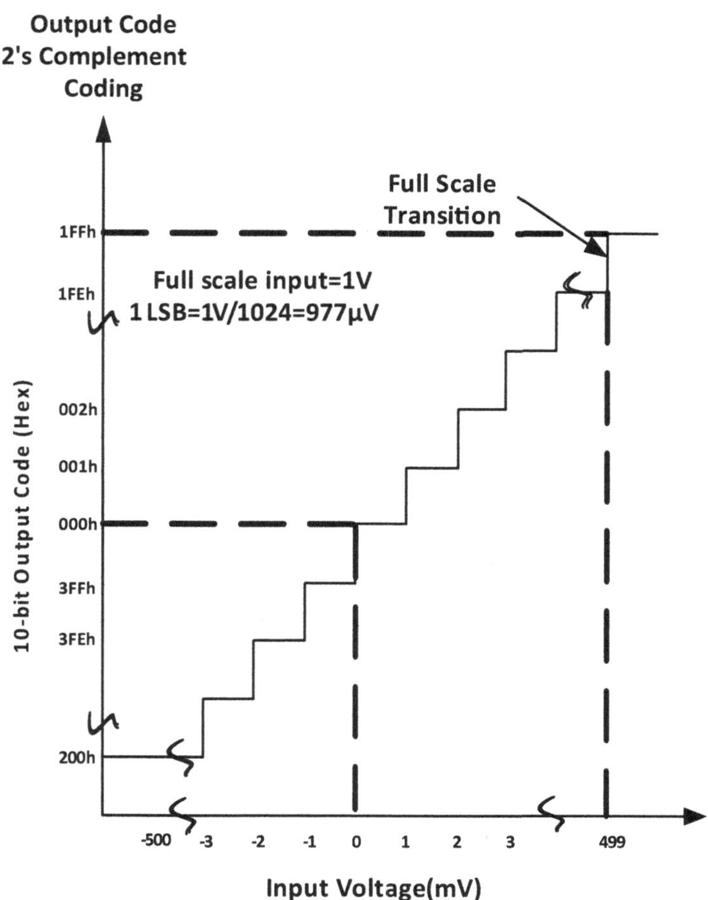

Fig. 16.4 Bipolar transfer function of SYSMON ADC

T=Temperature K (Kelvin)

When using the on-chip reference voltage, the transfer function for temperature sensor is as shown by (16.2):

$$T = \frac{\text{ADC} \times 501.3743}{2^{\text{bits}}} - 273.6777 \tag{16.2}$$

16.6.2 Power Supply Sensors

The SYSMON also includes on-chip sensors that allow monitoring of the device power supply voltages using the ADC. The sensors sample and attenuate the power supply voltages V_{USER} [3:0], V_{CCINT}, V_{CCAUX}, $V_{CC\_PSINTLP}$, $V_{CC\_PSINTFP}$, V_{CC_PSAUX}, and

V_{CCBRAM} on the package power supply balls. Equation (16.3) gives the power supply sensor transfer function after digitizing by the ADC. The power supply sensor can be used to measure voltages in the range 0 V to $V_{CCAUX}+3\,\%$ with a resolution of approximately 2.93 mV. The equation transfer function is related to the HP (High Performance) IO banks. The power supply measurement results are stored in the respective Status Registers:

$$\text{Voltage} = \frac{\text{ADCCode}}{1024} \times 3\text{V} \tag{16.3}$$

16.7 Controlling the SYSMON Operation

The SYSMON has a rich set of registers which can be accessed in three different mechanisms of interfaces (Fabric DRP access, I2C access, JTAG TAP access). Figure 16.5 shows the SYSMON register set. The access for up to 256 registers is allowed which are of 16-bit wide each, by any of the three interface mechanisms mentioned above. You need to follow the timing relation of different DRP ports while accessing these registers through fabric interface. The fabric register access is referred with respect to the *DCLK*.

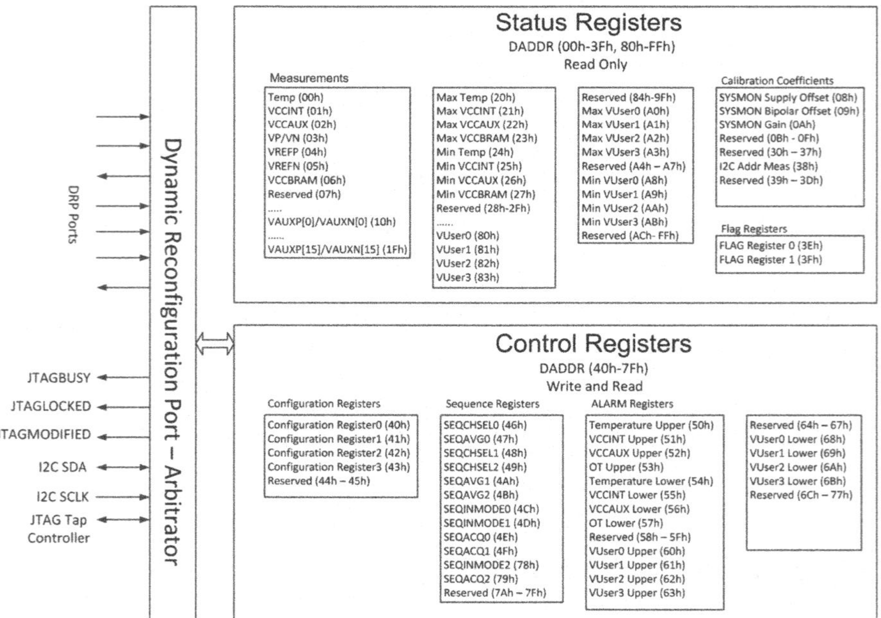

Fig. 16.5 SYSMONE1 register interface

16.7.1 SYSMON Control Registers

The *Control Registers* are used to configure the SYSMON operation. All the SYSMON functionality is controlled through these registers. These Control Registers are initialized using the SYSMON attributes when the SYSMON is instantiated in a design. This means that the SYSMON can be configured to start in a predefined mode after the FPGA configuration.

The Control Registers are further classified into:

- Configuration Registers (address range *0x40h* to *0x43h*)
- Sequence Registers (address range *0x46h* to *0x4Fh* and *0x78h* to *0x79h*)
- Alarm Registers (address range *0x50h* to *0x6Fh*)

The *Configuration Register* has bits associated with operating modes like *Sequence Mode*, *Single-Channel* or *External Multiplexer Mode* (*Auto Channel Sequencer*), *Continuous* or *Event Trigger Mode*, *Averaging Mode* on selected channel, *Channel Sequencing* operation, *Calibration settings*, etc.

Along with the *Control Register* configuration, there are multiple *Sequence Registers* available that need to be configured in order to help SYSMON to operate in the correct manner. In case of *Single-Channel* mode, the *Control Register* needs to be set to select only one of the available channels. In cases, when multiple channels need to be monitored, then *Auto Channel Sequencer Mode* is enabled. Based upon predefined sequence of channels defined in the *Channel Sequence Register* (*SEQCHSEL*), the sequencer automatically selects the next channel for conversion, sets the averaging (*SEQAVG*), configures the analog input channels (*SEQINMODE*), sets the required settling time for acquisition (*SEQACQ*), and stores the results in the *Status Registers*.

The ADC *Channel Averaging Registers* (*SEQAVG*) enable and disable the averaging of the channel data in a sequence. The result of a measurement on an averaged channel is generated by using 16 or 64 or 256 samples, which is controlled through *Configuration Register* bits. Offset correction enablement is also configurable option for ADC and supply sensors.

The SYSMON provides mechanism to raise the user intervention for any adverse condition occurring in the system using *Alarm Registers*. The *Alarm Registers* are used to set up the automatic alarms once the channel input signals crossover the limits set by you. The alarms are generated on 16-bit *ALM* port. You can program the alarm thresholds in the *Control Register* address range of *0x50h* to *0x6Bh*. The alarm for particular input will be set up both for the lower as well as higher limits. The alarms are reset when a subsequently measured value falls inside the threshold (min and max) ranges.

The on-chip temperature measurement is used for critical temperature warnings and also supports automatic shutdown of the FPGA device to help prevent the device from being permanently damaged. During very high temperature scenarios beyond 125 °C, the FPGA device is auto shut down; however this option needs to be enabled separately by you through configuration of the *Over Temperature* (*OT*) upper Alarm Register. The device auto-shutdown facility is disabled by default. During FPGA shutdown the SYSMON still maintains its data using the internal clock oscillator. The auto-shutdown facility is really useful as it prevents the device from getting permanently damaged. Once the on-chip temperature reduces, it is necessary to reconfigure the device for further usage.

User application can keep watch on the temperature alarm signals and should take the necessary action like turning on the cooling system, etc.

16.7.2 SYSMON Status Registers

The SYSMON *Status Registers* are *read-only* registers, which have the updates of all the measurements carried out by the SYSMON ADC. These registers can be accessed at the address range of *0x00h* to *0x3Fh* and *0x80h* to *0xBFh*. For each of the ADC capabilities, one individual register is provided. It includes parameters like on-chip temperature, different on-chip voltages (VCCINT, VCCAUX), external analog channel registers, etc.

There are two more set of *Status Registers* which are categorized in *MAX* and *MIN* type, which store the maximum and minimum values of these parameters since the FPGA is powered on or since last reset of SYSMON. The *MAX* and *MIN* set of registers are different than the regular set of registers which stores the latest ADC conversion values. The *Flag Registers* (address range *0x3Eh*, *0x3Fh*) are considered to be part of *Status Registers* with each bit indicating the status of parameters for various alarms and *Over Temperature* (*OT*).

SYSMON can be digitally calibrated to phase out any offset as well as the gain errors in ADC and power supply sensors using the *Calibration Registers* which are also part of the *Status Registers* (address range *0x08h* to *0x0Ah*). A built-in calibration function automatically calculates these coefficients.

16.8 Operating Modes of SYSMON

SYSMON provides access mechanism for a range of analog signals such as an on-chip temperature sensor, on-chip supply sensors, the dedicated analog input (*VP/VN*), the auxiliary analog inputs, and the user supplies. It provides multiple operating modes to select the analog signals used in a design. The default mode of SYSMON operation is restricted only for the on-chip sensors, which is available even when the SYSMON is not instantiated in your design. The default mode uses calibration and on-chip oscillators to automatically measure temperature, V_{CCINT}, V_{CCAUX}, and V_{CCBRAM}.

16.8.1 Single-Channel and Auto Channel Sequence Mode

The single-channel mode uses a *Configuration Register* (*0x41h*) to select the analog channel. By writing to the *Configuration Register*, a design can select different analog channels. In application where many channels need to be monitored, to avoid overhead on the processing system for reconfiguration of the *Control Register* each time, *Automatic Channel Sequence Mode* function can be used. The automatic channel sequencer sets up a range of predefined operating modes, where a number of channels (on-chip sensors and external inputs) are used. The sequencer automatically selects the next channel for conversion, sets the averaging, configures the analog input channels, sets the required settling time for acquisition, and stores the results in the Status Registers based on a once off setting. Averaging can also be selected independently for each channel in the sequence. The sequence mode is further categorized into *Single-Pass Mode* and *Continuous Sequence Mode*. The channel sequencer functionality is implemented using a set of 12 *Control Registers*. Section 16.7.1 contains more information about the different user configurable registers of the SYSMON.

16.8.2 External Multiplexer Mode

In some applications, where IO resources are limited and need to monitor several external analog inputs, in such cases the *External Multiplexer Mode* can be used. The external multiplexer can be connected to the dedicated analog inputs (like *VP*/*VN* ports) or one of the auxiliary analog inputs. Figure 16.6 shows how you can use *External Multiplexer* operation with *VP*/*VN* ports.

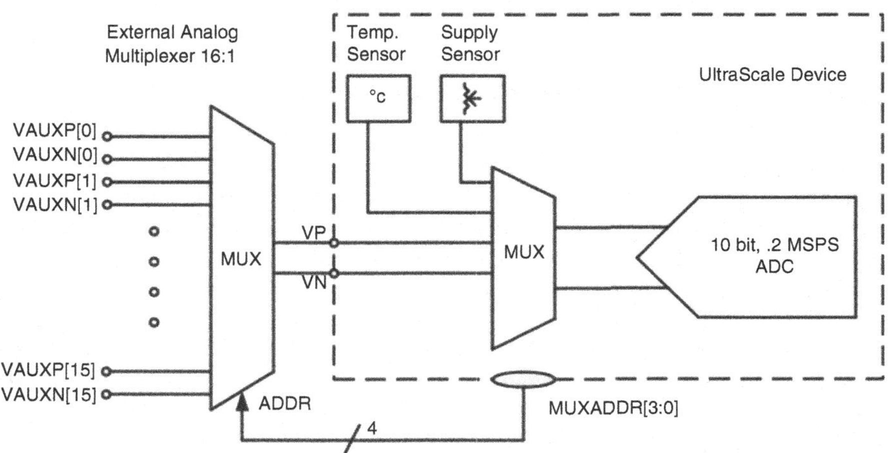

Fig. 16.6 External multiplexer mode

16.8.3 Automatic Alarms

The SYSMON also generates an alarm signal on the logic outputs *ALM*[*15:0*] when a sensor measurement exceeds some user-defined thresholds. The *Alarm Registers* are classified in to upper and lower alarm threshold control registers. The alarms are generated when the status register value for the corresponding recently measured channel goes outside the lower or upper limit mentioned in the *Alarm Threshold Control* registers (address range *0x50h* to *0x6Bh*). The alarms are suppressed automatically when the next new measurement of the channel falls within the range of upper and lower threshold registers.

16.8.4 Sampling Modes

The SYSMON has two modes of data sampling, namely, *Continuous Mode* and *Event-Driven Mode*. In *Continuous Mode*, the SYSMON ADC is busy doing the continuous conversion for the configured channel(s). A dedicated internal clock, namely, *ADCCLK*, is used to facilitate this conversion. The ADC takes around 26 *ADCCLK* clock cycles for any conversion. The maximum operating *ADCCLK* frequency is 5.2 MHz. The *ADCCLK* clock is dedicated only for the SYSMON ADC usage and it cannot be shared with other applications. For the SYSMON to operate in *Event-Driven Mode*, user application needs to provide one *DCLK*-wide active high pulse on the *CONVST* port of SYSMON. This pulse triggers the ADC to start the conversion of selected analog input. The *End of Sequence* (*EOS*) or *End of Conversion* (*EOC*) indicates that the current conversion cycle is over, and converted data is available in the respective Status Register.

The settling time of ADC decides the actual conversion rate. This flexibility helps the ADC to get 10-bit precise representation of analog conversion.

16.9 Using SYSMON in Standalone Mode

In *Standalone Mode*, SYSMON continuously monitors the on-chip sensors. Your application is no longer associated with the SYSMON results. Using the JTAG or I2C interface, you can also configure some of the external analog inputs. No instantiation is required to access the DRP interface over JTAG.

The JTAG as well as I2C access is slower as compared to the fabric DRP access. The SYSMON supports transfer up to 400 Kb/s, *Standard Mode* (SM) and *Fast Mode* (FM). For using the JTAG interface to access SYSMON, you need to be aware of the IEEE 1149.1 standard, while for I2C access, the knowledge of I2C 2-wire protocol is necessary.

The SYSMON could shut down the device operation if the junction temperature crosses 125 °C (Over Temperature (OT)), where it is still accessible only through JTAG access.

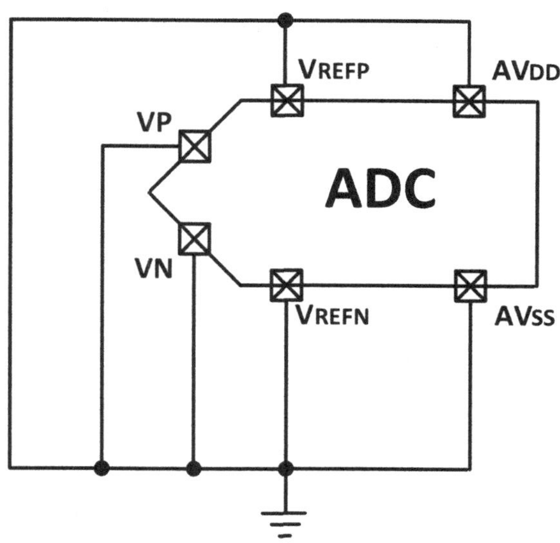

Fig. 16.7 Disabling the SYSMON

16.10 Disabling the SYSMON

Optionally, in order to save power further or in case your application does not require the inclusion of SYSMON, you can permanently disable it by connecting SYSMON's supply voltages to the ground. This is completed by connecting *AVDD, VREFP, VREPN*, and *AVSS* to ground reference. *VP* and *VN* also should be connected to *GND*. These groundings can be done while designing the PCB for your application. Alternately, XDC settings can be used during the system building process to disable SYSMON. As the SYSMON would not be available to monitor on-chip crucial parameters, you should carefully design your application to avoid any damage (like due to *Over Temperature*, etc.) to the FPGA. Disabling the SYSMON using external grounding of signals is shown in Fig. 16.7.

16.11 SYSMON Use Cases

The SYSMON can be found in various applications and use cases. Some of the applications are:

- Vehicle automation systems
- Food preservation systems
- Medical equipments
- Equipments used in harsh industrial atmosphere
- FPGA-centric critical systems

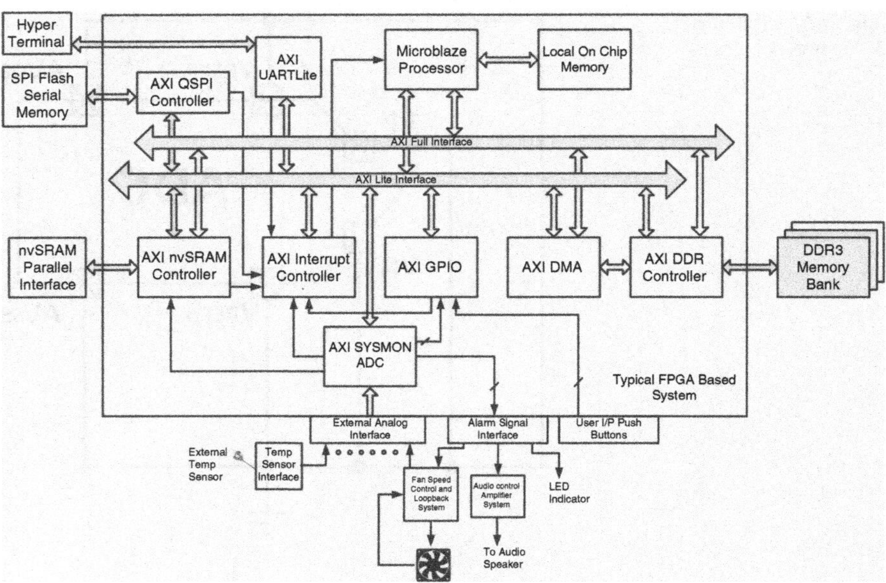

Fig. 16.8 FPGA-centric system based upon SYSMON functionality

Typical FPGA-based system is as shown in below Fig. 16.8.

In the FPGA-centric system shown in the Fig. 16.8, the SYSMON plays a crucial role of monitoring the internal parameters like on-chip temperature, voltage, etc. It also monitors the external parameters like on pressure, external temperature, incoming analog signals from cooling fan speed rotation control system, etc. Its *Alarm Signals* are connected to *Audio Control* System which controls the speaker activation. The *Alarm* Signals are also connected to a series of LEDs, which will be activated if any parameter under monitoring crosses the minimum or maximum range of user configurable parameters selected for monitoring purpose. The *Alarm* Signal also connects to cooling system. If the temperature goes beyond the limit, then the alarm signal automatically gets asserted and it controls the rotation speed of the fan. The set of Alarm Signals are connected internally to *GPIO* signals, which helps in monitoring these parameters when the system is in stable condition.

The interrupt signal from the *AXI*-based wrapper around SYSMON is connected to the dedicated input port of *Interrupt Controller* IP. The interrupt actions should be configured by the user based upon its severity. In critical cases like *Over Temperature (OT)* or if the on-chip voltage drops below minimum level, then the interrupt generated by the core will force the CPU to store the *present* status of its registers and application-related information in to nonvolatile SRAM memory. This can be done by using the *nvSRAM* controller (or equivalent *NAND Flash* controller if *NAND* memories are used as nonvolatile memories). The *nvSRAM* controller stores all the crucial information into *nvSRAM*, which can be used by the processor next time when the system reboots.

This system-level operation provides an example usage of SYSMON in FPGA-centric systems.

Chapter 17
Hardware Debug

Brad Fross

It is often necessary to debug FPGA designs in hardware, for several reasons:

- Problems are visible only when design is run in hardware at system speed.
- It is not feasible to re-create the failure in a simulation environment.
- It is faster to test the design in hardware than in a simulation or emulation environment.

This chapter discusses some of the advantages of debugging FPGA designs in hardware, how debugging complements other methods of verification and validation, and various techniques for getting the most out of debugging FPGA designs in hardware.

17.1 Debug Methodologies for FPGA Designs

17.1.1 Iterative Debug Methodology

Creating and debugging FPGA designs using an iterative design flow leverages one of the key advantages that FPGA devices have over ASICs and ASSPs: FPGAs are reprogrammable. Adding, modifying, and removing debug instrumentation are an integral part of the FPGA design cycle, as shown in Fig. 17.1. Due to their fixed nature, it is not possible to add, change, or remove debug instrumentation in ASIC/ASSP designs after fabrication.

While you can add debug instrumentation to ASIC/ASSP designs before tape out, it is typically only done at the block interface level or in key control sections of the design. It can be difficult to predict where bugs will pop up in a design, which

B. Fross (✉)
Xilinx, Longmont, CO, USA
e-mail: bkfross@ieee.org

S. Churiwala (ed.), *Designing with Xilinx® FPGAs*,
DOI 10.1007/978-3-319-42438-5_17

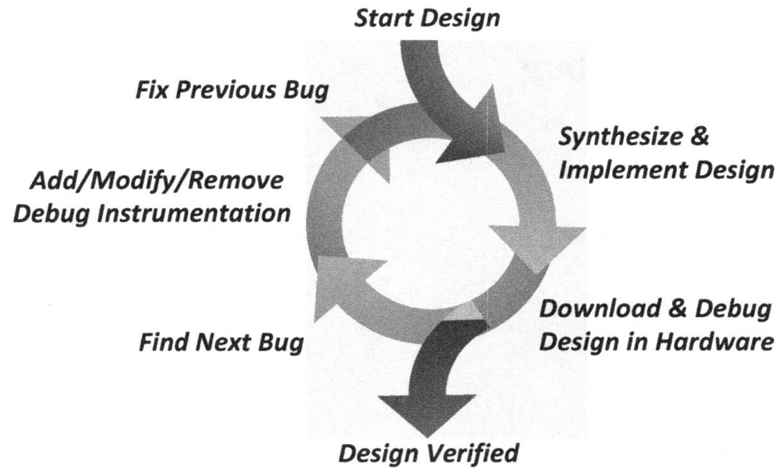

Fig. 17.1 FPGA design and debug cycle

means you might not be monitoring the appropriate part of the ASIC/ASSP design. In FPGA designs, you can add just the right amount of debug instrumentation to the appropriate part of the design to find the bugs and then remove the instrumentation when it is no longer needed.

17.1.2 Simulation vs. Debugging in Hardware

FPGAs are similar to ASICs/ASSPs in that simulation-based verification can be used to ensure the design meets the specification. The main advantage that simulation has over debugging in hardware is that simulation allows for full visibility of any node in the design. However, when debugging in hardware, the number of nodes that can be debugged in any given design iteration is limited to the amount of resources available to the debug instrumentation.

An example of debug instrumentation that is used to trigger on hardware events and capture data of interest is called the *Integrated Logic Analyzer* (*ILA*) debug core. The ILA IP core uses FPGA fabric resources to implement the trigger functions and it uses block RAM to store the captured data samples. Table 17.1 shows a chart of how many block RAM Tiles in the FPGA device are used for various levels of design debug visibility. The amount of *slice* logic used by the *ILA* core ranges from 0.81 to 2.61 % of a Kintex-7 XC7K480T device.

While simulation provides for increased debug node visibility and deeper capture trace, debugging in hardware has two distinct advantages over simulation:

- Faster test run times, typically at the speed of the system under test
- Testing in a real system environment rather than a simulation testbench

Table 17.1 Number of block RAM Tiles used by ILA cores of varying data dimensions

ILA core dimensions		Data depth					
		1024	2048	4096	8192	16,384	32,768
Data width	**32**	1	2	4	7.5	15	30
	64	2	4	7.5	14.5	29	58
	128	4	7.5	14.5	29	57.5	115
	256	7.5	14.5	29	57.5	114.5	229
	512	15	29	57.5	114.5	228.5	457
	1024	29.5	57.5	114.5	228.5	456.5	913

Simulation and hardware debugging are not mutually exclusive. In fact, it is often beneficial to use simulation to verify the functionality of a design before testing it out in hardware. This is especially true if the design consists of significantly new content that has not been previously verified. After verifying the design using simulation, you can debug in hardware to find issues that result from design integration or other system-level considerations, under the environment of real-world traffic patterns.

In some cases, it can be advantageous to skip simulation altogether and verify the design entirely in hardware. For example, in cases where designs that have been previously verified in simulation are undergoing small modifications or are being ported from one FPGA device family to another, you may go directly to hardware.

17.1.3 Debugging a Design That Meets Timing

Before debugging in hardware, make sure the design meets all timing constraint requirements. Debugging a design that does not meet timing is typically not a worthwhile endeavor since any misbehavior could easily be attributed to the failure to meet timing.

In addition to ensuring the design on its own meets timing, also ensure that the design including the debug instrumentation IP (such as the ILA core) meets timing as well. The ILA core uses FPGA device resources and can exhibit unexpected behavior if design does not meet timing.

The ILA core has a *clock* input that is used to synchronize the measurements to the design-under-test; therefore all design constraints related to that clock domain also apply to the ILA core. The ILA core also has its own design constraints that time the portions of the ILA core not related to the design clock domain. Once all timing constraints are applied correctly and are met, it is quite likely that any misbehaviors that are encountered are due to real functional issues as opposed to timing-related anomalies.

17.2 Instrumenting the Design for Debug

17.2.1 Choosing the Type of Debug Instrumentation

Deciding what type and amount of debug instrumentation to add to the design depends on two key factors:

- Type of issue being debugged
- Resources available for debug instrumentation

The types of design functionality issues can range from simple *status* and/or *control* issues to complex logic and/or system-level issues. The amount of resources available in the device (especially block RAM resources) can be the limiting factor in choosing the appropriate debug instrumentation IP. Table 17.2 shows how the Xilinx debug instrumentation options address various debug scenarios.

17.2.2 Choosing What Signals to Debug

It is important to consider two guidelines when choosing signals to debug:

- Select signals that will provide the necessary information to find and fix the bug without exceeding device resources
- Select signals that do not degrade the performance and/or functionality of the design

Table 17.2 Types of Xilinx debug instrumentation

Debug IP	Requires block RAM?	Debug scenarios
Integrated Logic Analyzer (ILA)	Yes[a]	• Useful when capturing samples in multiple consecutive clock cycles is required
		• Allows for complex triggering to find difficult-to-detect events
Virtual Input/ Output (VIO)	No	• Useful for low-bandwidth communication with design-under-test
		• Can replace or augment board-level control and status indicators such as buttons, LEDs, etc.
JTAG-to-AXI Master	Yes[b]	• Useful for reading/writing AXI-based peripherals
Integrated Bit Error Ratio Tester (IBERT)	Yes[c]	• Useful for debugging board-level signal integrity issues with high-speed serial I/O transceivers
		• Also used to determine transmit and receive margin of high-speed transceivers

[a]Amount of block RAM varies with IP port width and data depth parameters. See Table 17.1 for details
[b]JTAG-to-AXI Master uses two to six block RAM tiles, depending on IP parameters
[c]The number of block RAM depends on the number of transceiver QUADs available in the target device

When choosing what signals to debug, it is usually best to select signals that are driven by synchronous elements such as flip-flops, block RAMs, etc. The act of probing synchronous elements will not typically change the circuit unless the register would otherwise be combined into a primitive element (such as a block RAM output register or I/O block register). If you want to debug the outputs of combinational logic, it is important to consider how the act of probing the circuit will change its implementation by preventing the tools from optimizing it.

17.2.3 Choosing How to Add Debug Instrumentation

Along with deciding what signals to debug, it can be equally important to decide how to add debug instrumentation to a design. There are two methods for adding debug instrumentation:

- Source-level instantiation of debug cores
- Netlist-level insertion of debug cores

In the source-level instantiation method for adding debug instrumentation to a design, you directly instantiate the debug IP in the design source (e.g., HDL source code or *IP Integrator* block design). The debug IP is generated separately and can either be synthesized separately (i.e., *out of context*) or with the design-under-test (i.e., in context). You can add any of the debug cores described in Table 17.2 using the source-level instantiation method.

In the netlist-level insertion method, you add the debug instrumentation to the post-synthesis design netlist. You choose what signals to debug and how to debug them and then the Vivado software tools automatically insert the debug IP into the design-under-test netlist. You can add only the ILA debug core to the design using the netlist-level insertion method.

Some of the benefits for each method of adding debug instrumentation to a design are described in Table 17.3.

Table 17.3 Benefits of two methods for adding debug instrumentation to a design

Benefit	Source-level instantiation	Netlist-level insertion
Full correlation to source-level signals	Yes	No
Source code modification required	Yes	No[a]
Ease of probing across hierarchical boundaries	Low	High
Ease of adding, modifying, removing debug instrumentation	Moderate	High
Adding, modifying, or removing debug instrumentation requires design resynthesis	Yes	No
Adding, modifying, or removing debug instrumentation requires design re-implementation	Yes	Yes

[a]An optional step to improve the preservation of HDL signals during RTL synthesis is to use *MARK_DEBUG* or *DONT_TOUCH* properties on them

17.3 Interacting with Debug Instrumentation

Once you have added the debug instrumentation to the design, the design has been successfully implemented, and the design meets all timing constraints, it is time to program the design into the device-under-test and debug it. This section describes several ways to interact with debug instrumentation in a simple example design. Figure 17.2 shows the simple example design that contains a *MicroBlaze* microprocessor, a block RAM memory buffer, a *UART* peripheral, and a *GPIO* peripheral.

The example design shown in Fig. 17.2 has been instrumented with the following debug IP cores:

- An ILA core to monitor the AXI interfaces of the block RAM controller
- A *VIO* to monitor the inputs and outputs of the GPIO peripheral
- A JTAG-to-AXI Master to read/write the contents of the block RAM

The example design containing this debug instrumentation is shown in Fig. 17.3. The arrows indicate the new IP inserted due to the debug instrumentation.

17.3.1 Connecting to Hardware and Programming the Device

Before interacting with the debug cores in the design in hardware, you need to connect to the device-under-test via a JTAG cable and program the design into the device-under-test. In a typical lab environment, the JTAG cable for the target system is attached to the host machine that is running Vivado. In this case, clicking the *Auto Connect* toolbar button of the *Hardware window* in the *Hardware Manager* will connect to the locally attached JTAG cable, as shown in Fig. 17.4.

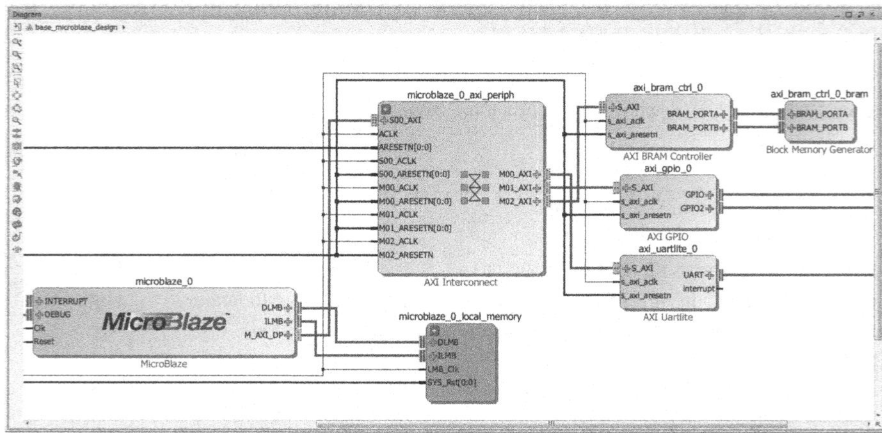

Fig. 17.2 Example design before adding debug instrumentation

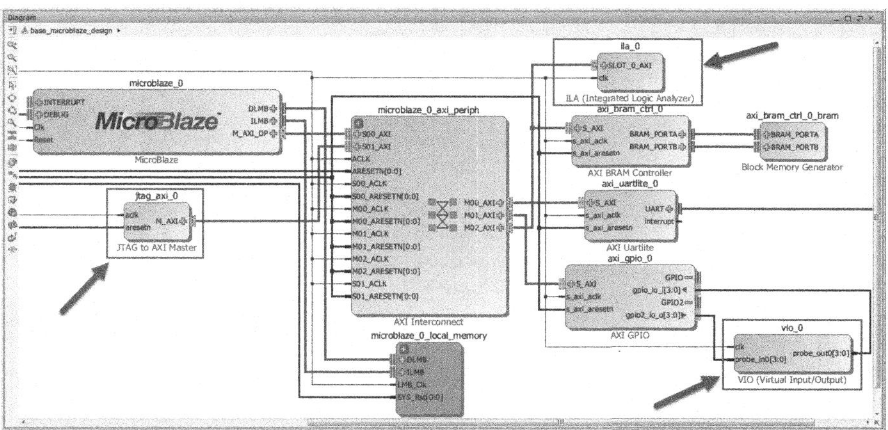

Fig. 17.3 Example design after adding debug instrumentation

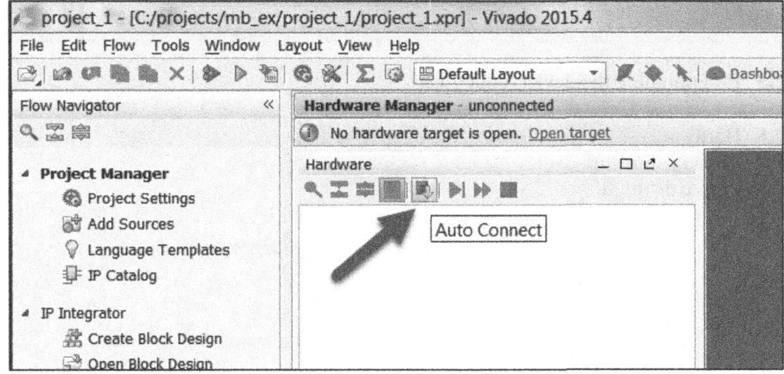

Fig. 17.4 Using Auto Connect to connect to a JTAG cable

Once the connection to the JTAG cable is established, Vivado scans the target system to detect the devices in the JTAG chain. After Vivado detects the *xc7k325t_0* FPGA device, the next step is to program the device with the bitstream file and debug probes file that correspond to the design in the current project, as shown in Fig. 17.5. The last step before debugging the design involves Vivado automatically detecting the debug cores in the device-under-test, as shown in Fig. 17.6.

17.3.2 Taking a Basic Measurement Using the ILA Core

The ILA core is very useful for triggering and capturing events as they occur in real time in the design-under-test. In the example design shown in Fig. 17.3, an ILA core is used to monitor transactions on the AXI interface of the block RAM controller peripheral.

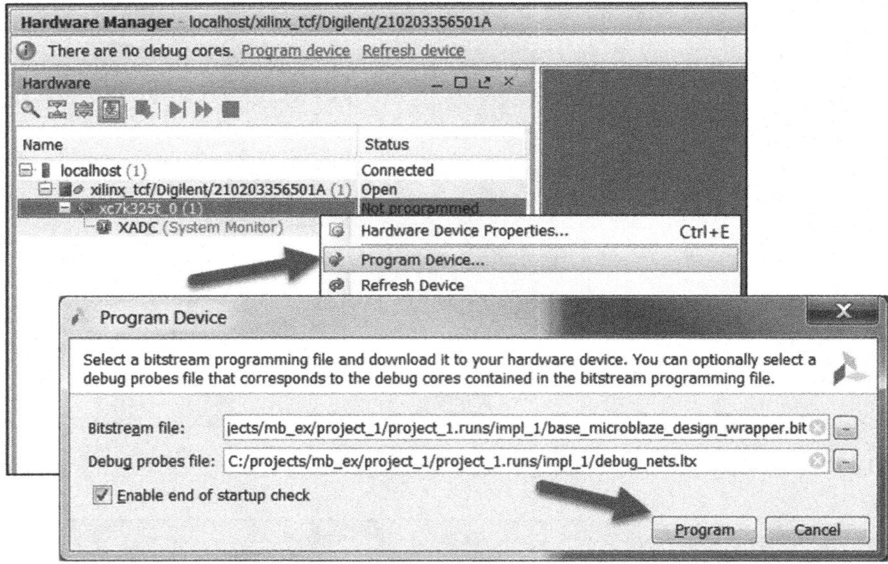

Fig. 17.5 Programming the device-under-test

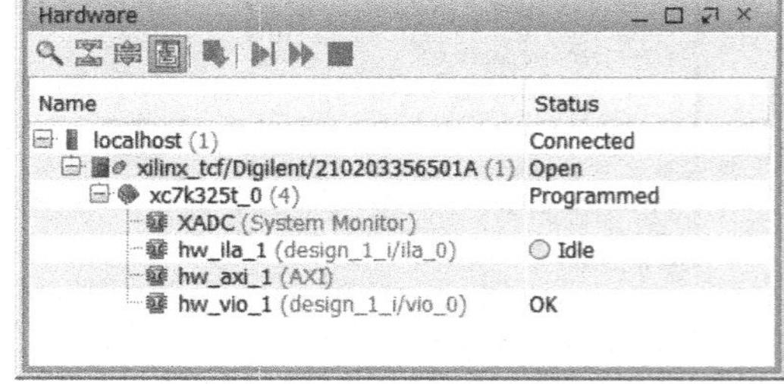

Fig. 17.6 Hardware window in Vivado showing device with three debug IP cores

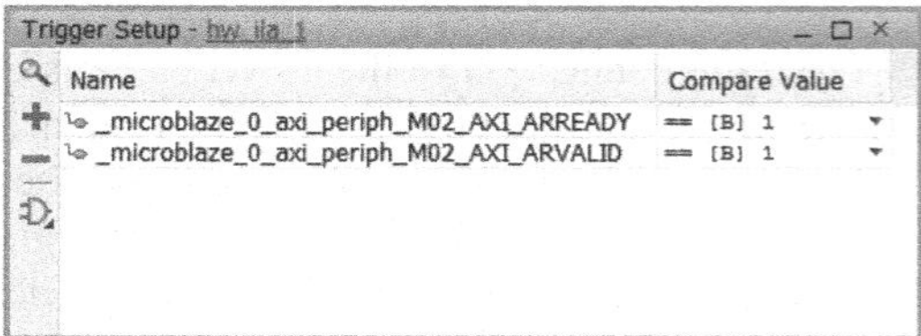

Fig. 17.7 ILA basic trigger setup for start of an AXI read transaction

A common measurement to take when monitoring an AXI interface is the start of *read* transaction which is signified by the *read address* acceptance event (*ARVALID = 1* and *ARREADY = 1*). The basic trigger setup of the ILA core can be used to take this measurement, as shown in Fig. 17.7.

After arming the ILA core to trigger on the start of an AXI read transaction, the ILA core waits for the trigger condition to occur. Once the trigger condition occurs, the captured data is uploaded and displayed in the waveform viewer, as shown in Fig. 17.8.

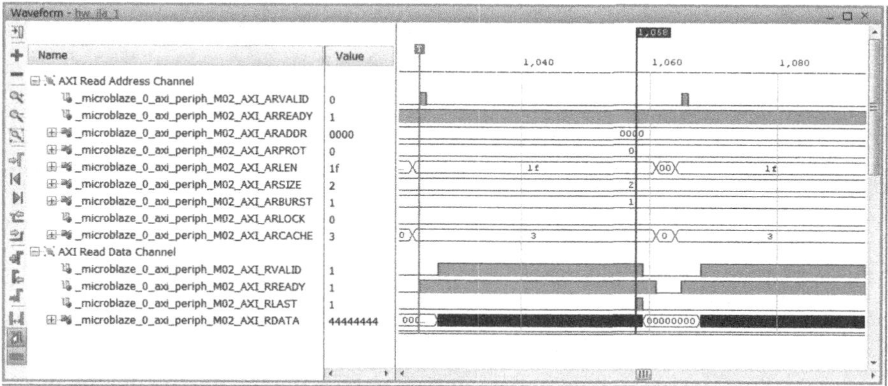

Fig. 17.8 Waveform showing captured data after first AXI read transaction

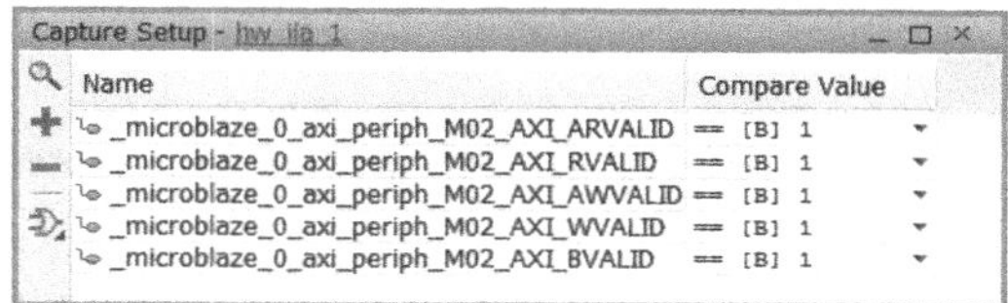

Fig. 17.9 ILA setup for capturing valid address and/or data

17.3.3 Maximizing the Usage of the ILA Core Data Capture Memory

The ILA core uses on-chip block RAM to store captured data samples. In its default setup, the ILA core captures a data sample every clock cycle following the trigger event. In many cases, it is desirable to only capture data samples that satisfy a particular capture condition. For instance, it might be desirable to only capture data samples when either the AXI *address* and/or *data* is valid. However, in the waveform shown in Fig. 17.8, many clock cycles worth of invalid *address* and/or *data* were captured following the burst of *read* transactions. By filling up the ILA's capture buffer with invalid data, subsequent bursts of valid data would be missed.

The following Boolean equation can be used as a capture *setup* condition (see Fig. 17.9) to only store only valid *address* and/or *data*:

$$
\begin{aligned}
Capture\,Setup\,Condition = \; & (ARVALID == 1)\,or \\
& (RVALID == 1)\,or \\
& (AWVALID == 1)\,or \\
& (WVALID == 1)\,or \\
& (BVALID == 1);
\end{aligned}
$$

Using the same basic trigger setup as shown in Fig. 17.7, the waveform in Fig. 17.10 shows how many more *read* transactions can be stored when only valid address and/or data cycles are captured.

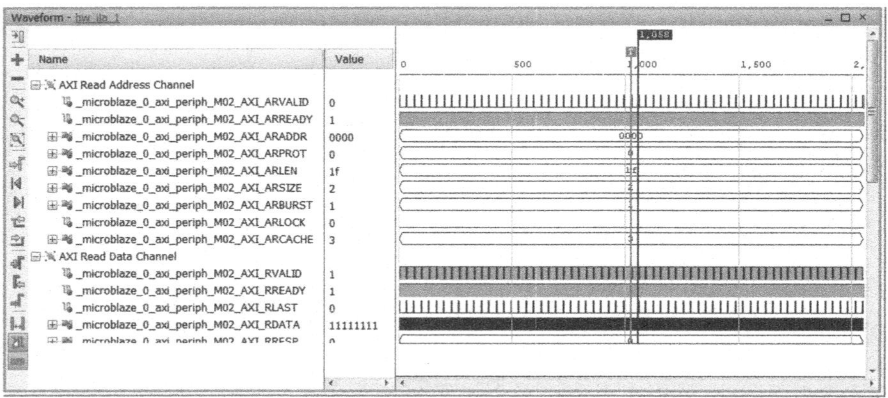

Fig. 17.10 ILA waveform showing only valid AXI address and/or data cycles

17.3.4 Taking an Advanced Measurement Using the ILA Core

Sometimes it is necessary to take a more advanced measurement with an ILA core than the basic trigger settings used in the previous section, as shown in Fig. 17.7. For instance, it is sometimes desirable to trigger when a particular AXI interface is idle for a certain number of clock cycles following a *read* transaction. This can be useful in detecting data transfer stalls or other throughput/latency issues.

You can use the ILA core's advanced trigger state machine feature to trigger on such an event, as shown in Fig. 17.11. You can optionally enable the ILA advanced trigger state machine feature at compile time and then use at run-time by selecting the *ADVANCED trigger* mode in the ILA trigger setup dashboard window. The waveform in Fig. 17.12 depicts the trigger occurring 1024 clock cycles after the last assertion of the *RLAST* signal.

17.3.5 Using JTAG-to-AXI Master to Access AXI-Based Registers

The *IP Integrator Block Design* AXI *Address Editor* shown in Fig. 17.13 has two AXI masters that are capable of initiating AXI transactions: a *MicroBlaze* microprocessor (*microblaze_0*) and a JTAG-to-AXI Master (*jtag_axi_0*). The JTAG-to-AXI Master debug IP core provides a means to access any register or memory that is in the AXI address map of the design. The JTAG-to-AXI Master debug IP core can be very useful for inspecting AXI-based memory contents or checking AXI-based status registers.

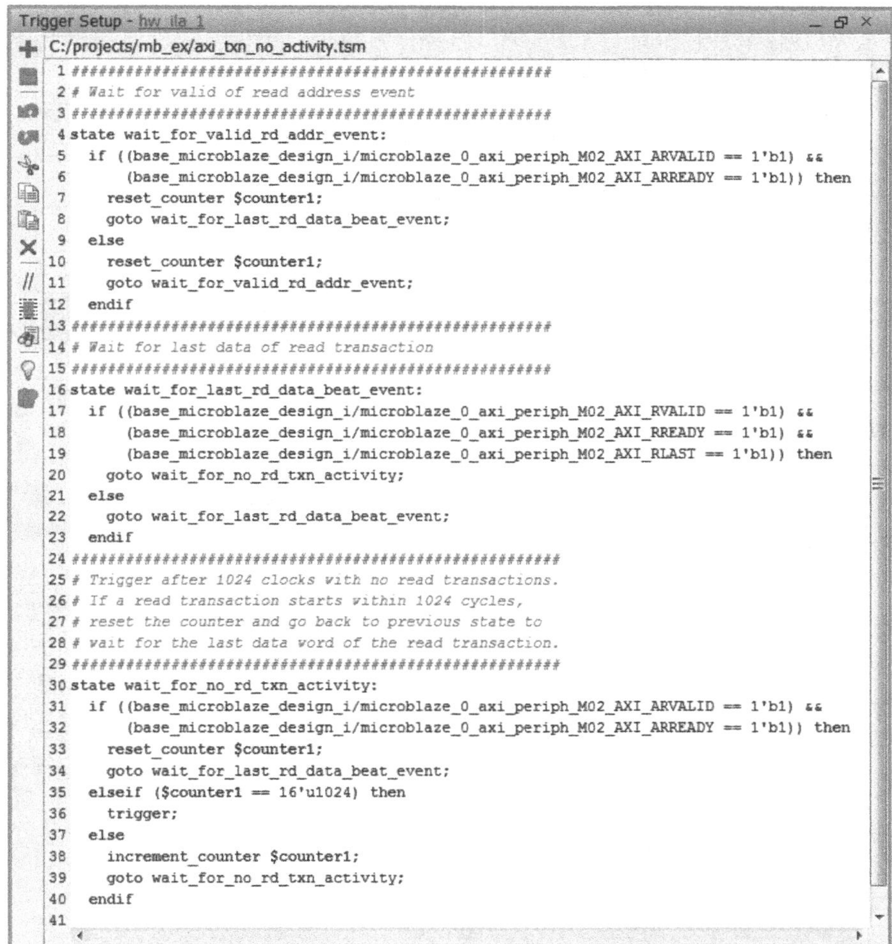

Trigger Setup - hw_ila_1

C:/projects/mb_ex/axi_txn_no_activity.tsm

```
########################################################
# Wait for valid of read address event
########################################################
state wait_for_valid_rd_addr_event:
  if ((base_microblaze_design_i/microblaze_0_axi_periph_M02_AXI_ARVALID == 1'b1) &&
      (base_microblaze_design_i/microblaze_0_axi_periph_M02_AXI_ARREADY == 1'b1)) then
    reset_counter $counter1;
    goto wait_for_last_rd_data_beat_event;
  else
    reset_counter $counter1;
    goto wait_for_valid_rd_addr_event;
  endif
########################################################
# Wait for last data of read transaction
########################################################
state wait_for_last_rd_data_beat_event:
  if ((base_microblaze_design_i/microblaze_0_axi_periph_M02_AXI_RVALID == 1'b1) &&
      (base_microblaze_design_i/microblaze_0_axi_periph_M02_AXI_RREADY == 1'b1) &&
      (base_microblaze_design_i/microblaze_0_axi_periph_M02_AXI_RLAST == 1'b1)) then
    goto wait_for_no_rd_txn_activity;
  else
    goto wait_for_last_rd_data_beat_event;
  endif
########################################################
# Trigger after 1024 clocks with no read transactions.
# If a read transaction starts within 1024 cycles,
# reset the counter and go back to previous state to
# wait for the last data word of the read transaction.
########################################################
state wait_for_no_rd_txn_activity:
  if ((base_microblaze_design_i/microblaze_0_axi_periph_M02_AXI_ARVALID == 1'b1) &&
      (base_microblaze_design_i/microblaze_0_axi_periph_M02_AXI_ARREADY == 1'b1)) then
    reset_counter $counter1;
    goto wait_for_last_rd_data_beat_event;
  elseif ($counter1 == 16'u1024) then
    trigger;
  else
    increment_counter $counter1;
    goto wait_for_no_rd_txn_activity;
  endif
```

Fig. 17.11 Triggering 1024 clock cycles after last AXI read transaction

Interacting with the JTAG-to-AXI Master IP involves two steps:

1. Create a transaction using the *create_hw_axi_txn* command.
2. Run the transaction created in step 1 using the *run_hw_axi* command.

Below is an example for creating and running a *read* and a *write* transaction, both of which are 32-word *bursts* starting at address *0xC000_0000* (the *axi_bram_ctrl_0* block RAM controller peripheral):

- Create and run a 32-word *burst read* transaction from address *C000000* and confirm the data is all *0*'s:

  create_hw_axi_txn rd [get_hw_axis hw_axi_1] -type read -address C0000000 -len 32
  run_hw_axi [get_hw_axi_txns rd]

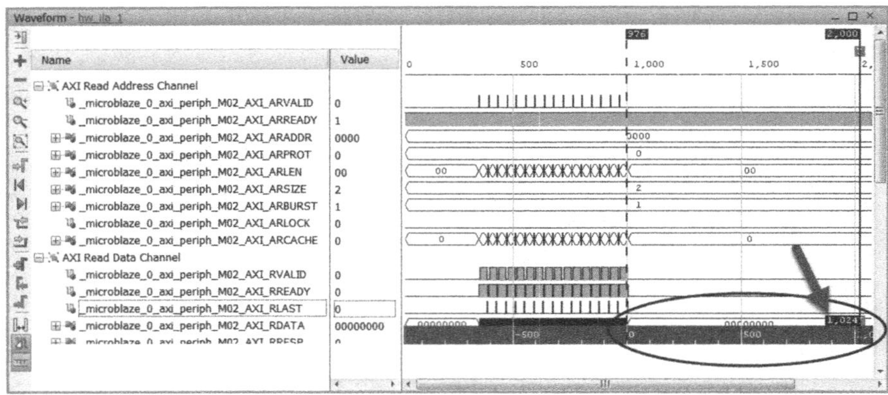

Fig. 17.12 Waveform showing trigger after 1024 idle cycles

Address Editor

Cell	Slave Interface	Base Name	Offset Address	Range	High Address
microblaze_0					
Data (32 address bits : 4G)					
microblaze_0_local_memory/dlmb_br...	SLMB	Mem	0x0000_0000	32K	0x0000_7FFF
axi_uartlite_0	S_AXI	Reg	0x4060_0000	64K	0x4060_FFFF
axi_gpio_0	S_AXI	Reg	0x4000_0000	64K	0x4000_FFFF
axi_bram_ctrl_0	S_AXI	Mem0	0xC000_0000	8K	0xC000_1FFF
Instruction (32 address bits : 4G)					
microblaze_0_local_memory/ilmb_bra...	SLMB	Mem	0x0000_0000	32K	0x0000_7FFF
jtag_axi_0					
Data (32 address bits : 4G)					
axi_bram_ctrl_0	S_AXI	Mem0	0xC000_0000	8K	0xC000_1FFF
axi_gpio_0	S_AXI	Reg	0x4000_0000	64K	0x4000_FFFF
axi_uartlite_0	S_AXI	Reg	0x4060_0000	64K	0x4060_FFFF

Fig. 17.13 Example design AXI address map

INFO: [Labtoolstcl 44-481] READ DATA is: 0000...

- Create and run a 32-word *burst write* transaction to address *C0000000* that repeats the four-word pattern of *11111111, 22222222, 33333333, 44444444*:

  create_hw_axi_txn wr [get_hw_axis hw_axi_1] -type write -address C0000000 -len 32 –data {44444444_33333333_22222222_11111111}
  run_hw_axi [get_hw_axi_txns wr]
  INFO: [Labtoolstcl 44-481] WRITE DATA is: 4444...

- Rerun the *read* transaction:

  run_hw_axi [get_hw_axi_txns rd]
  INFO: [Labtoolstcl 44-481] READ DATA is: 4444...

- Confirm the data at address *C0000000* is the same pattern that was previously written:

  report_hw_axi_txn [get_hw_axi_txns rd]

c0000000 11111111 22222222
c0000008 33333333 44444444
...
c0000078 33333333 44444444

This sequence of AXI *read* and *write* transactions confirms the block RAM controller peripheral is working as expected.

17.3.6 Using Virtual Input/Output to Debug Design in Hardware

The Virtual Input/Output (*VIO*) debug IP core is useful for representing status indicators and high-level controls such as LEDs and pushbuttons. You can use the VIO core in situations where the hardware is not physically accessible or there are not sufficient interactive controls on the hardware platform.

The example design in Fig. 17.3 shows how a VIO core can be used to monitor the inputs to and outputs from an AXI *GPIO* (General Purpose I/O) peripheral. The VIO dashboard in the Vivado tool can be used to show the value of the *GPIO* outputs (which are inputs to the VIO), as shown in Fig. 17.14.

The *JTAG-to-AXI Master* IP can also be used to write a nonzero value to the GPIO outputs:

- Create a transaction to write *0x0000000F* to the *GPIO* output register at address *0x40000008*:

  create_hw_axi_txn gpio_f [get_hw_axis hw_axi_1] -type write -address 40000008 -data {0000000F}

- Run the transaction:

  run_hw_axi [get_hw_axi_txn gpio_f]

Note that the value of the VIO inputs (GPIO outputs) changed from all zeroes to all ones, as shown in Fig. 17.15).

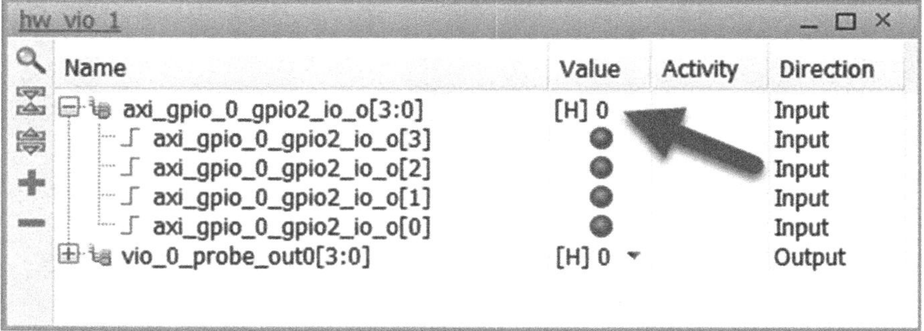

Fig. 17.14 VIO dashboard showing inputs are all zeroes

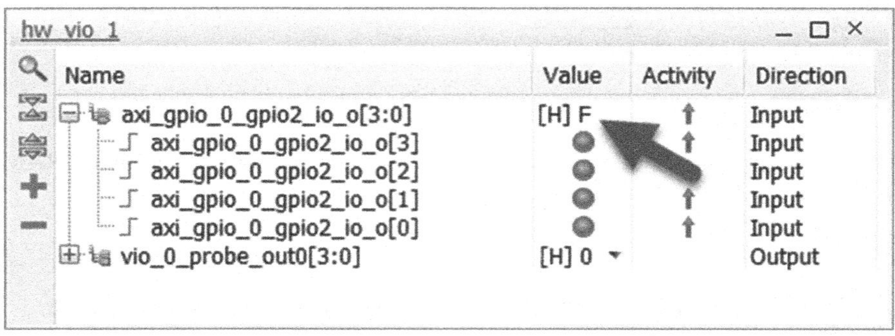

Fig. 17.15 VIO dashboard showing inputs are all ones

17.4 Board-Level Debugging

In addition to debugging system designs internal to the FPGA or the PL portion of an *MPSoC* device, you can also use the Vivado hardware debug tools to debug board-level issues. Below are some of the board-level debug features included in the Vivado tool:

- Debugging high-speed serial I/O signal integrity issues and measuring the transmitter and receiver margin using the *Integrated Bit Error Ratio Tester* (*IBERT*) debug feature
- Debugging external memory calibration issues and measuring read and write margin using the Calibration Debug feature of External Memory Controller
- Measuring on-chip temperature and voltage sensor values using the System Monitor feature

Usage of external memory, transceiver, and System Monitor is described in Chaps. 4, 5 and 16, respectively.

Chapter 18
Emulation Using FPGAs

Paresh K. Joshi

18.1 Introduction to Emulation

For the purpose of this chapter, we will use emulation to include prototyping also—since underlying challenges and methodologies are common. We read about simulators in Chap. 11. An emulator is a *simulation-specific* hardware, which is capable of retaining the parallelism of the blocks of the design, thereby significantly improving the speed of execution.

Depending on the capabilities of the emulator, you can get very close to your design environment. Since emulators are dedicated hardware, the speed advantage is obtained at the cost of observability and controllability. Emulation also needs additional setup, which is what this chapter is mostly about. In an ideal scenario, the emulator must support all the features of simulation at a speed and cost advantage.

18.1.1 Types of Emulators

1. *Array of simulation-specific processors* (*Cadence Palladium series*): Array of processors whose instruction set and software is tailored to simulation tasks. One set of such arrays is called a *board*. Each processor on the board can simulate millions of gates in parallel. Furthermore, each processor on the board talks to other processors via a fixed (specific) protocol.
2. *Array of FPGAs (Synopsys ZeBu series):* Array of FPGAs. Each FPGA can have mapped gates programmed into it. Each FPGA in the Array usually has dedicated wiring with other FPGAs.

P.K. Joshi (✉)
Intel Mobile Communications, Bangalore, Karnataka, India
e-mail: paresh.k.joshi@intel.com

S. Churiwala (ed.), *Designing with Xilinx® FPGAs*,
DOI 10.1007/978-3-319-42438-5_18

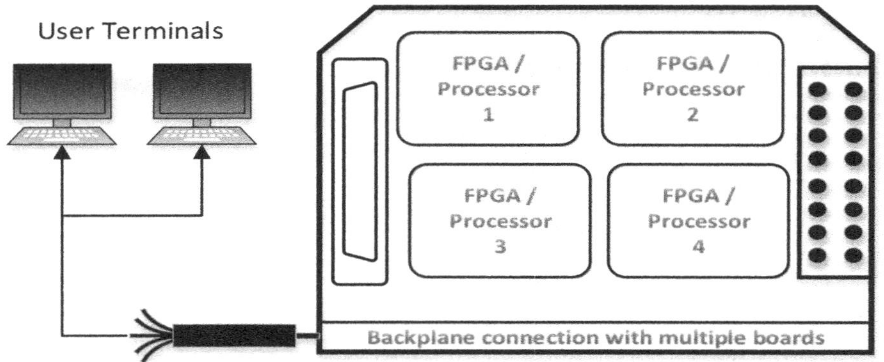

Fig. 18.1 Cascading 4 processors/FPGAs to build a larger emulation system

3. *A hybrid array of both simulation-specific processors and FPGAs (Mentor Veloce series).*

For large designs boards in an emulator can be cascaded. To better utilize the components in the emulator, there are partitions possible which enable multiple users to simultaneously access the resources of the emulator.

Since emulators comprise of hardware components, it is possible to connect the emulator to real external targets like JTAG, UART, QSPI, I2C, etc. The JTAG and UART are used by the software team to do hardware-software co-design and debug at the *programmers view* level.

Figure 18.1 illustrates an FPGA or processor array-based emulator system with multiple user terminals, standard connectors, IOs, and a backplane to cascade multiple such boards. Multiple users can then use the emulator boards for improving resource utilization.

18.1.2 Uses of Emulation/Prototyping

Substitute for simulation: This is the most obvious usage. In practice, however, emulation is resorted to only after the RTL design reaches a certain level of maturity. A not-so-mature RTL design will find iterative debug to be difficult, due to limited observability and controllability of emulation.

Enabling pre-silicon software development: Once the RTL is reasonably mature, the software teams can use the emulator for developing BOOTROM, Software (UBOOT, Linux, Android, RTOS, UEFI), Device Drivers (BSP), etc. Doing so provides several months of lead time to the software teams. This enables the software components to be available and ready for use, immediately after the device silicon is available.

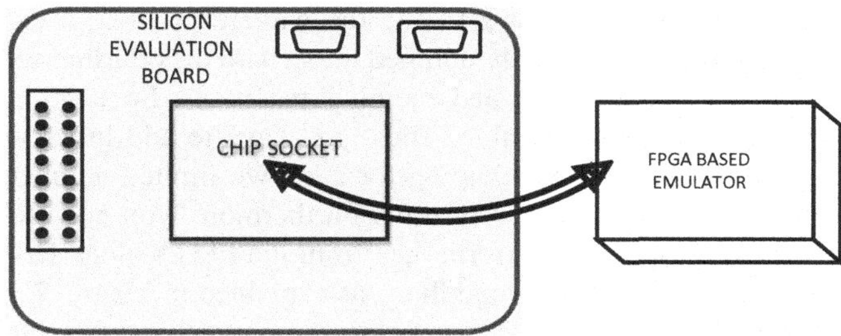

Fig. 18.2 Silicon Evaluation Board with a socket being interposed with FPGA-based emulator

Place-holder for actual silicon: The first silicon bring-up team designs an evaluation board with sockets for the device. Before the actual silicon is available, the emulator can behave as a prototype and fit into the socket using a plug-in board. The evaluation board along with silicon bring-up test cases can be run on the system as shown in Fig. 18.2.

18.2 Emulation Using FPGAs

System designers and prototyping teams have been using FPGAs to their benefit. FPGA tools are available to provide RTL to FPGA mapping. If you have a prototyping environment, the additional activities for going to emulation include:

1. Creation of a synthesizable and reconfigurable testbench.
2. Addition of instrumentation into design for advance debug purposes.
3. Mapping of complex design blocks like IOs, SERDES, DSP blocks, and block RAMs to the FPGA.
4. Remapping of complex clocking structure of the device to the FPGA-based PLLs and clock controllers.
5. Mapping of design IOs to the FPGA IOs to obtain connectivity to the external targets (JTAG, UART, etc.).
6. For designs which require multiple FPGAs:
   (a) *Logic Partitioning*: Partitioning of the design into chunks of logic to fit into individual FPGAs. This depends on the size of the design and the size of placeable gates on the FPGA. The logic and memory closely associated with the said logic are grouped together into pieces which fit on the same FPGA.
   (b) *Pin Partitioning*: Partitioning of the design with appropriate pin count across FPGAs. This depends on the hardware board design and is usually fixed for a particular board.

The additional activities for going to emulation from a simulation setup include:

1. *Observability:* Simulation allows to see the waveforms for all signals at all times. The waveforms are directly dumped into a hard disk during runtime. In an FPGA, there are limited logic and memory resources. So complete runtime waveform dumping is not possible. Thus, you have to add instrumentation to *trigger* the start of waveform dumping for a known limited number of signals and for a known limited amount of time. Furthermore, you need to build in a mechanism to retrieve the waveform data from the FPGA block RAMs. Xilinx Vivado provides ILA core for doing this—as explained in Chap. 17.
2. *Controllability*: For some tests, a specific pin (say: *reset*) may need to be kept at a desired value for a specific duration. In simulation you can force the signal then release it. A similar ability needs to be provided when doing emulation using FPGAs. Xilinx Vivado provides VIO.
3. *Memory initialization*: The DUV usually contains BOOTROM which needs to be programmed (preloaded) with the appropriate bitmapped code. The testbench could have other memory models of flash, DDR, etc. In the simulation environment, the memory load (*$readmemb*) and dump can be used. A similar ability is required for emulation using FPGAs.

Xilinx FPGAs and the Vivado tool set provide the methods and means to make all of the above possible.

18.3 Challenges in Emulation Using FPGAs

The basic challenge is to stitch the hardware, the tool software, and the RTL-mapping flow with the evaluation board and components. This section breaks up the challenge into multiple parts and sections. Section 18.4 then explains on how to deal with these challenges.

18.3.1 Design Logic and Memory Size

The engineering choice is to use one FPGA which fits the design. However, sometimes the DUV may be bigger than the largest FPGA available. Even otherwise, sometimes fitting the DUV into two smaller FPGAs is cheaper than using the largest FPGA available. If the design is skewed toward huge memory blocks, the FPGA tools can map parts of unmapped logic on the FPGA tile for memory blocks. For an emulator using FPGAs, (since the testbench is embedded into the FPGA) large memories like flash, DDR pose mapping problems. In such scenarios the emulator is fitted with large external memories which are then remodeled to behave like flash and DDR. Note that this remodeling is done through custom instrumentation insertion prior to using Vivado P&R tools.

18.3.2 Design Pin Count

The FPGA (or an array of FPGAs) must be able to support the relevant pin count of the device being emulated. In general, for emulation purposes a synthesizable testbench is used, indicating that there are fewer external connections. In certain cases, flash memory can be real components on the board which are then pinned-out to the board.

18.3.3 Clocking

Clocking between FPGAs and ASIC/ASSP is different. In an ASIC/ASSP there could be many hundreds of clock domains with multiple PLLs embedded. Each root clock derived from a PLL can have multiple secondary clock generation logic (say for dividing clocks, test clocking). Furthermore, sets of flip-flops or registers in the design can have clock-gating circuit implemented as part of power-reduction techniques.

FPGAs usually have a limited number of PLLs and a limited number of balanced clock channels incident upon a larger cluster of flip-flops. The challenge is to straighten up the ASIC clocks to map it easily onto the FPGA clocking.

18.3.4 RTL Constructs and Remodeling

Several RTL constructs are not FPGA friendly. These need to be modeled appropriately for FPGA. The remodeling has to be done without modifying the functionality. A module RTL makes it easier and scalable since there is a great usage of common cells in the design.

18.3.4.1 IO Pads Modeling

IO pads typically have tristate functionality. Usually, these IOs of the DUV are connected to the BFMs in the testbench. Recent FPGAs do not have built-in tristate gates. For FPGA usage, you need to remodel the tristates as shown by the example in Fig. 18.3. The Xilinx ISE/Vivado toolset automatically transforms internal tristates into logic elements.

18.3.4.2 ADC Module Modeling

For a module with analog behavior (e.g., ADC/DAC), you need to appropriately model to ensure that its boundary talking to the digital side of the design is clean. For example, an ADC module can easily be modeled with a memory and digital

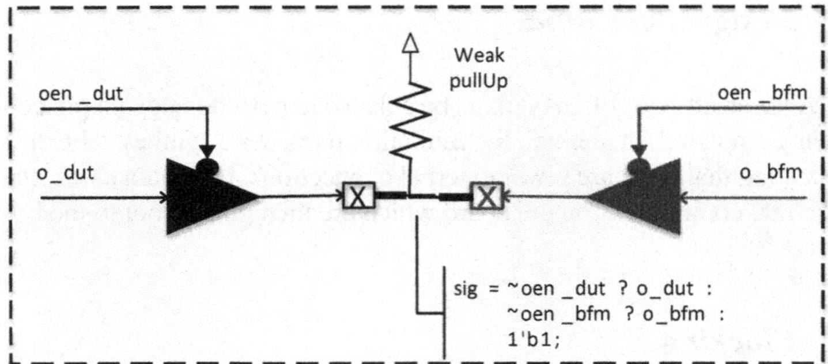

Fig. 18.3 Remodeling of typical IO connectivity within testbench between DUV and BFM

output. The memory can be preloaded with the kind of analog behavior we expect out of the design. Alternatively, an ADC can be placed on the FPGA board and the digital output can be used as an input to the design. If the ADC module is deeply embedded into the DUV, you need to bring out the wires from the embedded hierarchies onto the top level of the testbench.

For Xilinx FPGAs you can use the SYSMON module (explained in Chap. 16). However, you still need to take care of:

- Performance of the SYSMON for emulator clocking
- The analog stimulus to be fed to the SYSMON
- The appropriate remodeling of the ADC to instantiate the SYSMON into it

18.3.4.3 Memory Modeling

Typically the RTL has memories which are either ASIC technology memories or modeled as a memory array. Also, the RTL memory model could have test logic embedded into it. Remodeling memories for FPGA is typically a four-step process.

1. Identify the memories in the design. If the memories belong to the same technology node, then the entity is usually identical except for the *address* and *data* width. Sometimes, there might be variants (e.g., byte-wise write).
2. Remodel the memory component with an equivalent FPGA friendly construct. If you are not interested in test logic, they could be tied to their disabled state. This remodeled memory component is then verified to be true using simulation. If the memory needs to have user-defined preloading or dynamic preloading, then explicit instrumentation needs to be added.
3. One level of FPGA synthesis and run is carried out to flush out the flow.
4. Create a scriptware to convert all the flavors of *data* and *address* widths.

Steps (2), (3), and (4) are true for all types of remodeling done at RTL level, but it deserves a special mention for memories since there are many types.

18.3.4.4 Standard Cells Modeling

It is best to have synthesizable view of the technology standard cells in the design. Most technology libraries provide the synthesizable view of standard cells.

18.3.4.5 Inferred Components Modeling

Some RTL descriptions infer multipliers, dividers, special Register Files, FIFOs, etc., during the ASIC synthesis flows. These components use compiled models/ descriptions for simulation. Such components will end up as being unresolved. A way to resolve this problem is to actually do an ASIC synthesis and use the verilog equivalent for the said component. Thus:

FPGA RTL view = synthesized netlist from ASIC tool + the synthesizable RTL view of technology std-cell

18.3.5 FPGA Board Design

The FPGA-based emulation system is very much dependent on the FPGA board design. In particular, the number of FPGAs in the array, the capacity of each FPGA in the array, the external memory connected (for modeling large memories, for dynamic waveform dumping, and for using memory as Look Up Table for large pieces of logic with huge fan-in cones), and the external connectors, switches, GPIOs, and LEDs are provided. Its levels of complexity are higher to move from one FPGA-based emulator to another than it is to move across simulators from different vendors. The basic complexity is due to the use of hardware for emulation and so it is fixed. This complexity makes it difficult to make sound design and financial decisions for the right choice of FPGA-based emulators. FPGA vendors provide a chart with logic gate count estimates, IOs, memory blocks, SERDES blocks, and DSP blocks within the FPGA.

18.4 General Methodology

In this section we provide some known recipes to the challenges explained in Sect. 18.3. The recipes below would help design teams to realize their own FPGA-based emulator. We have assumed (by this chapter, toward the end of the book) a basic understanding of FPGA-based design.

Note that you should perform RTL to RTL Logic Equivalence Check after any RTL transformation.

18.4.1 RTL-Related Transformations

PLLs: All technology ASIC libraries contain PLLs. Each PLL consists of basic *reference clock in*, *clock out*, with pins indicating the multiplier factor in terms of *Numerator* and *Denominator* values. These have to be mapped to the equivalent PLLs in the selected FPGA. The methodology used is to keep the ASIC PLL entity identical but to instantiate the FPGA clocking resource in place. If the PLL has multiple clock outputs, the same are also remapped to the FPGA.

Clock Dividers: If there are dividers in the design, then it is appropriate to remove the divider circuits and replace them with the FPGA clock resource outputs as defined in the MMCM clock tile.

It would be useful to maintain a table similar to Table 18.1.

In the Table 18.1, for (#2) and (#3), the clock frequencies are the same, i.e., 20 MHz. It would be worthwhile to investigate from an ASIC clocking point of view, if it is possible to use the same PLL output of 20 MHz driving the clock end points of both (#2) and (#3). If the clocks are of the same frequency, but asynchronous to each other, it would be OK to reduce the use of a PLL and free up routing resources and reduce complexity of mapping to the FPGA.

Programmable Clock Dividers: Usually there is a use of Programmable Clock Dividers to select a baud rate as it is in the case of UART. In such cases, reconfigurable registers of the ASIC need to be remapped to the Dynamic Reconfiguration Data Input of the Clocking tile. Most emulation designers would put the dynamic reconfiguration data input as part of the instrumentation in the testbench, so that they have better control over the clock.

Clock Gating Cells: Integrated clock gating cells are instantiated by the RTL designer to enable dynamic power reduction. This can be a problem with FPGAs which can get resource limited if there are too many clock gating cells in the design. A solution is to do a tool-based or hand-scripted transformation to the clock gating cells. A typical example is provided in Fig. 18.4.

Table 18.1 Mapping of ASIC clock frequencies to FPGA clocks

#	ASIC clock	ASIC freq	FPGA clock resource	FPGA freq	Comments
1	Clock.A	400 MHz	PLL1.CLKOUT0	40 MHz	All clock scaled as div by 10
2	Clock.B	200 MHz	PLL2.CLKOUT1	20 MHz	
3	Clock.A.div2	200 MHz	PLL1.CLKOUT1	20 MHz	A divider in the path of ClockA is remapped to a clock output synchronous to div2 of the PLL1.CLKOUT0
4	Clock.A.div8	50 MHz	PLL1.CLKOUT2	5 MHz	Div8 of the PLL1. CLKOUT0

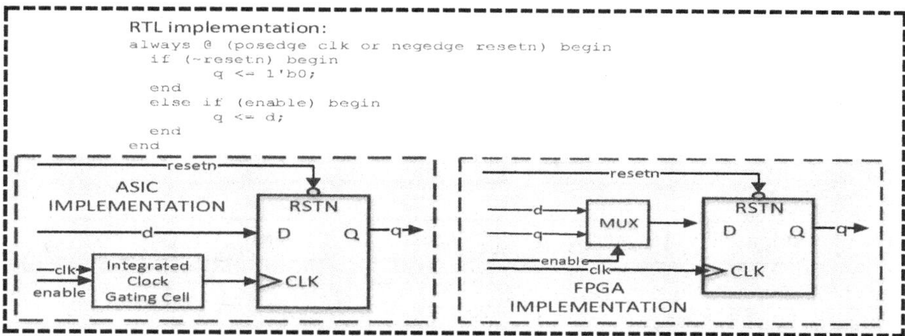

Fig. 18.4 Typical ASIC and FPGA implementation for a clock gating cell

18.4.2 Multiple FPGA Specific (The Partitioning Problem)

Now that the individual pieces of your RTL have been readied for FPGA-based emulation, the next level of complexity comes if the design cannot be mapped on one FPGA. For a particular design, it might not fit into a single FPGA, due to either of the following:

- Design logic size exceeding the logic that can be mapped onto the FPGA.
- Design logic could be mapped, but it could not be routed.
- Design logic was mapped and routed, but design has more memory than the block RAMs on the FPGA.
- Design ran out of IO that could be appropriately mapped on the FPGA.

Irrespective of the situation leading to the use of multiple FPGAs, all of the above need to be resolved on a per FPGA basis on a MultiFPGA emulation system. To start with, get a gate, memory, and pin count estimate for the big blocks in the design. Also, assume that each FPGA may be about 60 % utilized to begin with. Typically, most big IPs would fall within 5 ~ 6 sub-hierarchical levels of logic. This exercise would give a rough estimate of the number of FPGAs required to fit the design and testbench.

The exercise is iterative. Start with partitioning through the most constrained of the three resources (gate count, pin count, memory) and then affect the grouping changes to see if the other constraints can also fit. Figure 18.5 depicts the hierarchical view of the DUV and the testbench BFM components and the Table 18.2 the tabular view of the same. Both these views (hierarchical and tabular) help in converging to the right partitioning between multiple FPGAs.

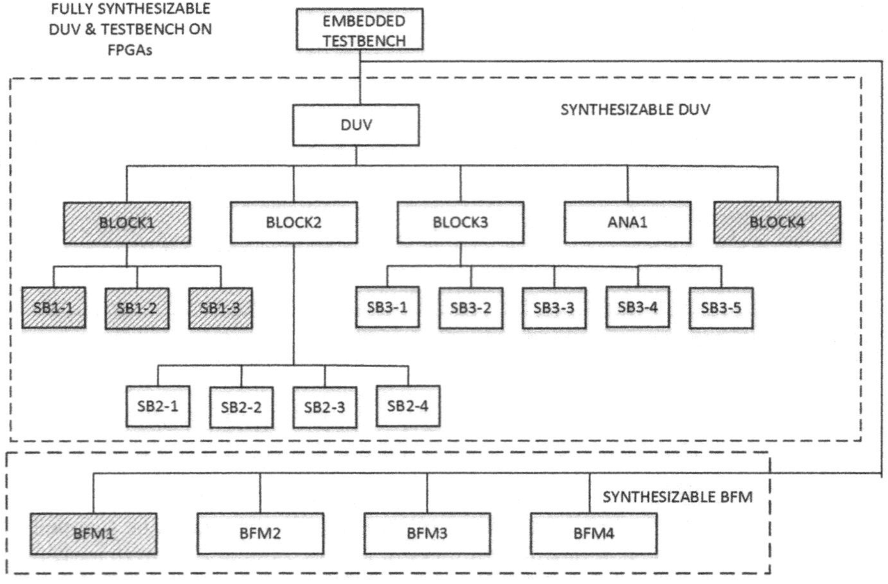

Fig. 18.5 Hierarchical view for embedded synthesizable testbench with DUV and BFM

Table 18.2 FPGA view for the embedded synthesizable testbench with DUV and BFM

subHier Level	ModName	GateCount	PinCount	TotalMemory	Estimate FPGA
1	tb_top	250	200	4 Mbits	
2	tb_top.BFM1	12M	100	200 Kbits	FPGA1
2	tb_top.BFM2	24M	50	100 Kbits	FPGA2
2	tb_top.BFM3	14M	125	250 Kbits	FPGA3
2	tb_top.DUV	200M	350	3.5 Mbits	
3	tb_top.DUV.BLOCK1	75M	450		FPGA1
3	tb_top.DUV.BLOCK2	80M			FPGA2
3	tb_top.DUV.BLOCK3	35M			FPGA3
3	tb_top.DUV.BLOCK4	5M			FPGA1
3	tb_top.DUV.ANA1	5M			FPGA3

18.4.2.1 Partitioning Gate Count Challenge

Once the gross level partitioning is known through analytical method as per Table 18.2, we need to get the same implemented. There are tools which can read in the RTL files and then dump out a regrouped file. Such grouping would result in new hierarchical tables being generated, as shown in Table 18.3.

For this example, considering per FPGA gate count of ~100M gates, Table 18.3 shows that FPGA3 is OK, but FPGA1 and FPGA2 are likely challenges to the P&R

Table 18.3 Sorted list of hierarchies on per FPGA basis

subHier Level	ModName	GateCount	PinCount	TotalMemory	Estimate FPGA
1	tb_top	250M	200	4 Mbits	
2	FPGA1.BFM1	12M	100	200 Kbits	**FPGA1**
2	FPGA1.BLOCK1	75M			**FPGA1**
2	FPGA1.BLOCK4	5M			**FPGA1**
2	FPGA2.BFM2	24M	50	100 Kbits	FPGA2
2	FPGA2.BLOCK2	80M			FPGA2
2	FPGA3.BFM3	14M	125	250 Kbits	**FPGA3**
2	FPGA3.BLOCK3	35M			**FPGA3**
2	FPGA3.ANA1	5M			**FPGA3**

Table 18.4 Actual partitioned pin count vs. available connections between FPGAs

F1 <--> F2	F1 <--> F3	F1 <--> F4	F2 <--> F3	F2 <--> F4	F3 <--> F4
PF12	PF13	PF14	PF23	PF24	PF34
IPF12	IPF13	NA	IPF23	NA	NA

stage. These considerations and iterations go on until there is sufficient convergence. Table 18.3 is deficient in terms of pin count and memory as it is for illustration purpose only.

However, since the module BLOCK2 and BFM2 are closely knit with each other, there could be pin count challenge if some readjustments of modules of BLOCK2 are done onto FPGA3 which seems to be least constrained.

18.4.2.2 Partitioning Pin Count

The MultiFPGA board usually has fixed pin count which can be summarized in a template table as in Table 18.4.

In Table 18.4 PF12 are the physical IO pins that are available between FPGA1 and FPGA2 (F1 <--> F2) on the FPGA board.

In Table 18.4 we have a Not Applicable (NA) if the particular FPGA is not used in the implementation. The implemented pin count across the FPGAs (IPF) should be less than the provisioned pin count across the FPGAs (PF). Thus, the pin count criteria can be converged when IPF12 < PF12 and so on.

If the pin count criteria are not satisfied, you could resort to pin muxing for the IO. This means that another utility RTL needs to be added to send multiple bits of data over a single IO from one FPGA to another. This utility RTL is inserted prior to the pin-multiplexed IO. Figure 18.6 shows the circuit for the utility RTL on the FPGAs for pin multiplexing. There are three main operations done:

- Load: convert from parallel to serial.
- Shift: shift the serial data from FPGA2FPGA.

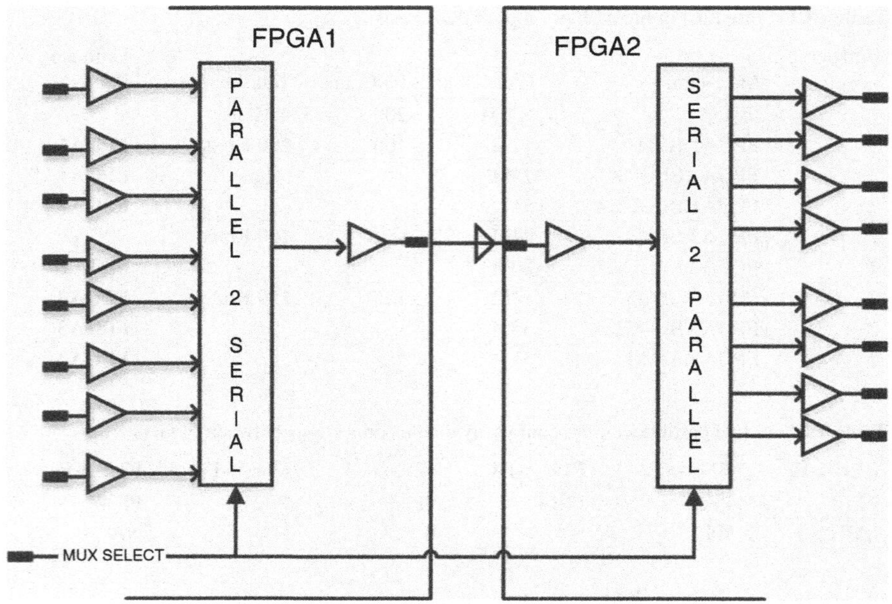

Fig. 18.6 Pin muxing for IOs over two FPGAs

- Restore: convert serial data back to parallel.

EDA Tools like Certify™ from Synopsys® form a major backbone to enablement of this convergence.

18.4.2.3 Using SERDES Lanes

It is also possible to use the FPGA SERDES Lanes as an extension to the pin multiplexing. SERDES provides a convenient *serializer* and *deserializer* over a two-wire network, which can transmit and receive data Gbps (Giga bits per second) range. The SERDES lanes are useful in converting FPGA2FPGA IOs into serial, sending it across at high speed and reconstructing the same at the other end.

18.4.2.4 Handling Clocks Over Multiple FPGAs

As soon as we move into using multiple FPGAs, the clocking complexity increases. One way is to see each hop or evaluation as a phase (a dedicated time slot) and increase the emulation clock period accordingly. This means that the performance of the emulator drops every time there is a signal hop.

18.5 Instrumenting

There are ways of achieving some degree of controllability and observability on an FPGA-based emulator, albeit at the cost of performance, logic area, and memory requirements. A general observation is that about 10~40% (depending on design specifics) of the design overhead on an emulator is attributed to addition of instrumentation for controllability and observability. At each step of the instrumentation addition, exercise care to maintain the equivalence of the design.

Let us assume that the emulator adds an instrumentation port (say Instrumentation JTAG or iJTAG) through which it can carry out the functions of observability and controllability to the design. This instrumentation port provides an interface to the user using a host computer. Figure 18.7 logically explains the two ports needed for an emulator. Modern emulators like Synopsys ZeBu use the PCIe as an instrumentation port.

18.5.1 Ability to Stop and Start the Emulation

The emulator start-stop is affected by the clocking. If the clock to the logic blocks does not tick, the emulator is in *stop* state. The instrumentation needed to achieve the purpose are:

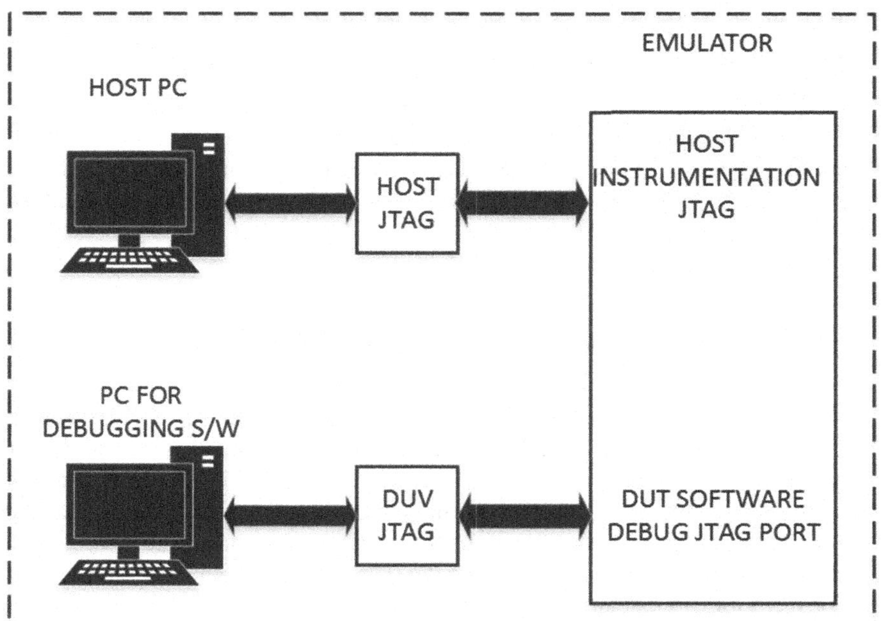

Fig. 18.7 Instrumentation (iJTAG) port connecting host computer and the emulator

1. Create a set of clock gates in instrumentation through the use of the *BUFGCE*, *BUFGMUX*, etc. The *BUFGCE* is used for *Enable*. The *BUFGMUX* is a mux between *instrumentation* mode and *functional* mode.
2. Create a set of counters, preferably one per primary clock. It should be possible to start, stop, and free run the counter. A set of count comparators, then could gate the clock to the functional logic blocks. Through the iJTAG one can write into these instrumentation registers which control the counters and clocks.
3. Using similar control instrumentation, you can also have some DUV internal signals *trigger* or *stop* the emulator clocks.

18.5.2 General Observability of Signals and Registers in the Design

The RTL synthesis process for FPGA optimizes out intermediate combinatorial logic signals. This scenario is in contrast with "array of processor"-based emulators, where each node can be maintained within the processor database.

- For the registers, using the iJTAG port, and decoding logic-related instrumentation, it is possible to have full controllability and observability. Figure 18.8 gives a feel of the instrumentation to be added for a register (flip-flop).
- For intermediate signals (part of combinatorial logic), a monitor flop and control mux can be added to gain controllability and observability.

There are multiple methods to enable these instrumentations:

- Modify the RTL to add pragmas known to Xilinx Vivado tool suite.
- Use a netlist editor tool post functional synthesis.
- Use a dedicated vendor tool for instrumentation insertion. Example Synopsys ZeBu tool suite does a seamless instrumentation insertion tailored to the ZeBu FPGA-based emulator.

18.5.3 Instrumentation for DUV Internal Memory

Often, it is needed to preload internal ROM and SRAMs with the executable code. The *C* program for the application is compiled, linked, and loaded into internal memories. The intent is to release the CPU reset and expect the CPU to execute the code and data loaded into the respective memories. Instrumentation can be added and accessed using the iJTAG as per the Fig. 18.8 even for memories. Note that the functional ROMs can also be preloaded using the iJTAG after instrumentation insertion.

For memories like dual-port memories, the port which has both write and read ports is chosen for instrumentation. Table 18.5 indicates the typical instrumentation that needs to be inserted for commonly used memories within the DUV.

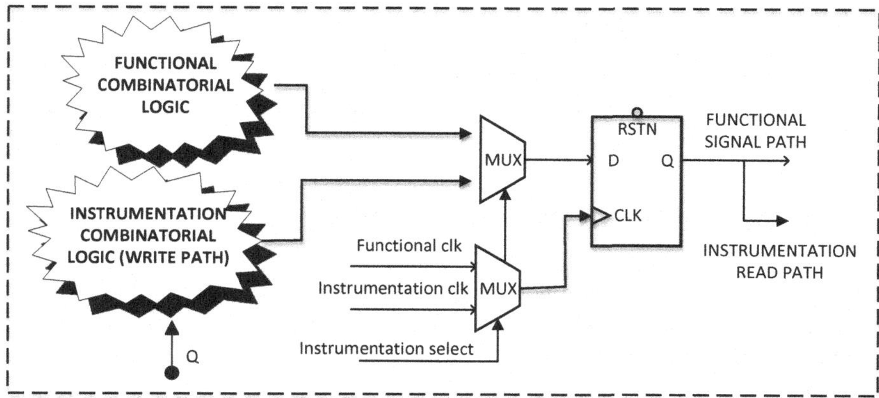

Fig. 18.8 Control and observability for registers using instrumented logic

Table 18.5 Typical instrumentation needs for memories

Memory	Functional	Instrumentation
ROM	Read only	(a) Clock muxing
		(b) Write port addition
		(c) Address and data line muxing
Single-port (SP) RAM	Read and write	(a) Clock muxing
		(b) Address and data line muxing
		(c) Write/read control signal muxing
Dual-port (DP) RAM	Different types	(a) Clock muxing on any one Write Port
	(a) 1 W, 1R	(b) Insertion of read port instrumentation for the write port (if it does not exist)
	(b) 1 W&R, 1R	(c) Address and data line muxing (for instrumented port)
	(c) 1 W&R, 1W&R	(d) Write/read control signal muxing

If the SP/DP RAM has bit- or byte-wise write and read control (functionally strobed lanes), then the instrumentation is suitably adjusted so that all the byte lanes are affected during memory load and dump through iJTAG.

The typical sequence for the usage would be:

1. Stop all the clocks to the emulator. This is through iJTAG-based instrumentation register configuration.
2. Preload the memories using external iJTAG:
   (a) Glitch-free selection of the clock to point to iJTAG_TCK.
   (b) Select the memory to be preloaded.
   (c) Preload the memory with the (address, value) pairs.
3. Apply reset to the DUV.
4. Start the clocks to the emulator.

5. Release reset to the DUV.
6. Expect the design to run the test (application).
7. Stop all the clocks to the emulator.
8. Read the memory (address, value) pairs, and store it to a file on host machine.

18.5.4 Adding Signal Observability (Waveforms)

Observing waveforms is an important part of the debug process and this feature is integral to any emulator. With regard to waveform, there are a few key concepts that need to be put in place as below:

1. *Signal List*: List of signals and buses (full hierarchical names) that you want to be added into the debug waveform.
2. *Trigger Signals and Trigger Expression*: A set of *Trigger* signals and the Boolean expression which would control the start and stop of the waveform capture.
3. *Trace Depth*: The maximum number of *waveform samples* that can be taken using the appropriate sampling clock.
4. *Trace Window*: The period of time when the waveform samples are captured. You can also have a circular trace buffer, allowing for a % trigger start, i.e., the trace starts $x\%$ prior to the actual trigger event and lasts up to $(100-x)\%$ after the trigger event. One can also define a pre-trigger percent or a post trigger percent based on this as is indicated by Fig. 18.9.

Chapter 17 explains various debug cores provided by Xilinx that can be used to capture waveforms. However, often, for deeper level of debug, the ILA is not sufficient, and at times the *Signal List* can span multiple FPGAs. To address this problem, emulators usually have their own external SRAM/DDR memory which can go up to 128 GB to enable deep trace. Intuitively, one can see that the instrumentation needed for this feature is huge. Some basic components are listed in Table 18.6.

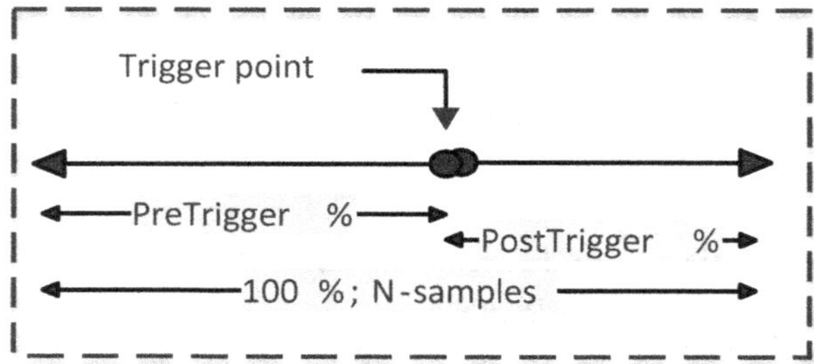

Fig. 18.9 Illustration of Trigger Point and “pre- and post trigger percent”

Table 18.6 Instrumentation components for waveforms using external memory

Instrumentation component	Usage
DDR Memory	The waveform samples would be written in the DDR memory. The samples are then read back and stored onto a host file
DDR Controller	To adhere to the DDR protocol for writing and reading the DDR memory
Signal Funnel	An instrumentation logic which converts (packs) the Signal List into chunks of data for writing and reading to the DDR memory
Instrumentation clock	Addition of an instrumentation clock, which is typically 1× or 2× the frequency of the sampled signals
Optional instrumentation CPU subsystem (iCPU)	The triggering, capturing of set of signals would need an instrumentation CPU to control the flow. The CPU would control the traces written to the DDR, and can also help in reading the traces and formatting for waveform generation by appropriate usage of iJTAG (host port connection)
	If an iCPU is being added, it can also be configured to enable other instrumentation tasks including complex clock management for starting and stopping the emulator

Chapter 19
Partial Reconfiguration and Hierarchical Design

Amr Monawir

Partial Reconfiguration takes advantage of hierarchical design capabilities available in the Xilinx Vivado Design Suite. This chapter describes the various designs that can benefit from the use of Partial Reconfiguration, as well as the key concepts and design considerations for Partial Reconfiguration and the other hierarchical design flows available.

19.1 Partial Reconfiguration

FPGA technology provides the flexibility of programming and reprogramming a device with a modified design in the field without the need to go through re-fabrication. Partial Reconfiguration takes this one step further, allowing the dynamic modification of *part* of an operating FPGA design without impacting the rest of the design.

19.1.1 Applications

Any system with functions that can be time-multiplexed stands to benefit from taking advantage of Partial Reconfiguration. Using Partial Reconfiguration allows functions to be switched on hardware, similar to a microprocessor's ability to switch between tasks in software.

A. Monawir (✉)
Xilinx Ireland, Dublin, Ireland
e-mail: aye20@hotmail.com

S. Churiwala (ed.), *Designing with Xilinx® FPGAs*,
DOI 10.1007/978-3-319-42438-5_19

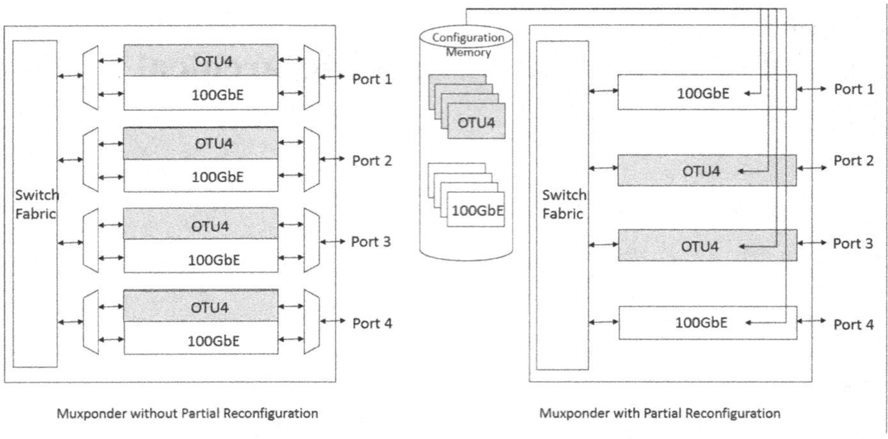

Fig. 19.1 100G Muxponder design implemented without and with partial reconfiguration

19.1.1.1 Multi-protocol Networking

In *optical transport network* (*OTN*), client side ports need to support multiple interface protocols. To ensure this, every possible interface protocol has to be independently implemented for each port. This is resource intensive and inefficient, especially considering that only one protocol will be used per port at any one time. Partial Reconfiguration allows the different protocols for each port to be dynamically loaded on demand. This removes redundant logic and provides a more efficient use of resource to implement the same functionality. Figure 19.1 shows the same 100G Muxponder system implemented with and without Partial Reconfiguration.

19.1.1.2 SW-Controlled HW Coprocessing

Hardware coprocessing is achieved by off-loading compute-intensive functions from the central processor to a coprocessor or dedicated hardware, which executes the function with lower power and latency. Image and video coprocessing is a typical example of this approach.

Having dedicated hardware for each function is an inefficient use of resources. Partial Reconfiguration allows a library of hardware functions to be partially reconfigured onto the same set of FPGA resources as and when required. Figure 19.2 gives an example of a processor system, with an array of dedicated hardware coprocessing functions, implemented with and without the use of Partial Reconfiguration.

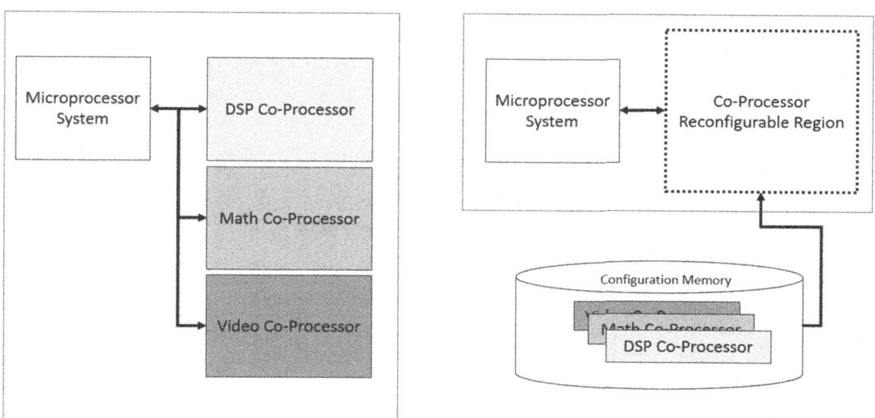

Fig. 19.2 A microprocessor system with dedicated hardware coprocessors implemented without partial reconfiguration on the left and with partial reconfiguration on the right

19.1.1.3 Security and Encryption

Encryption and public-private asymmetric key cryptography are widely used as a means of protecting sensitive or proprietary data. Partial Reconfiguration can be combined with asymmetric key cryptography to provide secure encrypted bitstream or data transfer. The encryption key generation and/or decryption engine on the FPGA is part of the initial or *static* part of the design. The encrypted *partial* bitstream containing the proprietary data is then sent to the decryption engine, decrypted inside the FPGA, and programmed via the *internal configuration access port* (*ICAP*), thus ensuring that the *partial* bitstream is never unencrypted outside the FPGA.

Figure 19.3 gives an example of how a decryption engine can be used in conjunction with Partial Reconfiguration.

19.1.2 Key Concepts

All Partial Reconfiguration designs consist of three basic parts. The *Static* is the portion of the design that does not change and is expected to continue to function at all times. The *Reconfigurable Partition* is the instance or level of hierarchy within which multiple *Reconfigurable Modules* are defined and implemented. Each *Reconfigurable Module* represents one of the time-multiplexed functions that will be switched in and out of the FPGA (Fig. 19.4).

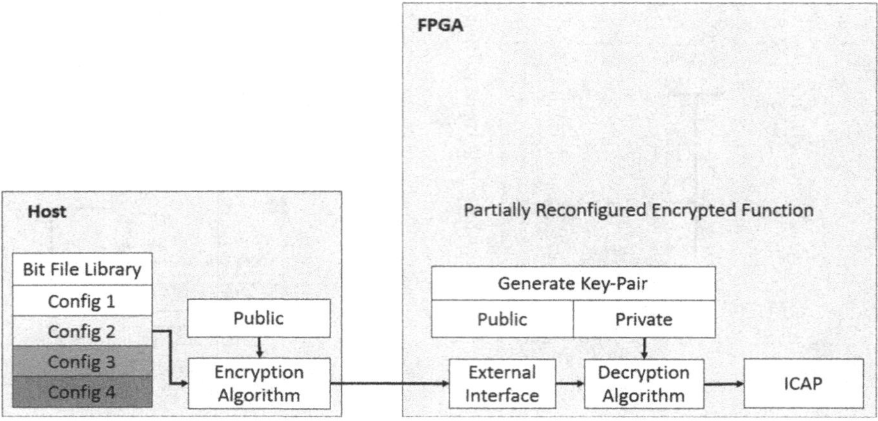

Fig. 19.3 Delivery of encrypted bitstreams using partial reconfiguration

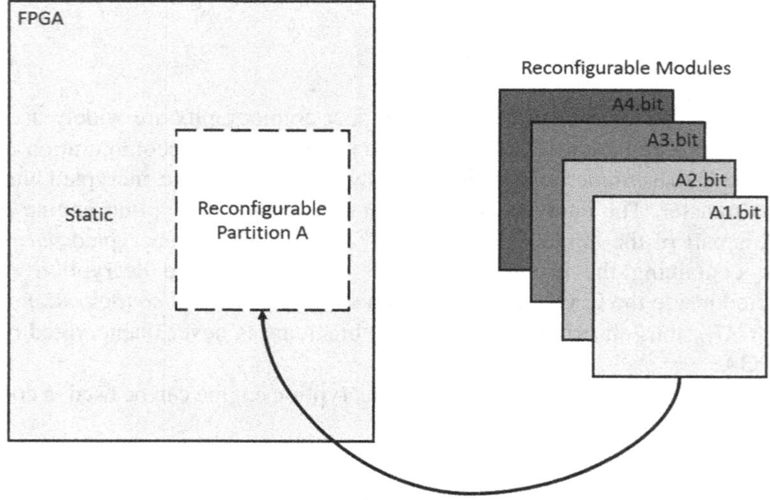

Fig. 19.4 Basic partial reconfiguration concept and terminology

Partial Reconfiguration designs can contain one or more *Reconfigurable Partitions*, each of which must occupy a mutually exclusive physical area of the FPGA. The physical area for a given *Reconfigurable Partition* must contain the aggregated resources required to individually implement each of the *Reconfigurable Modules* associated with it. The resource types and granularity of the physical area within the FPGA that can be reconfigured at any given time vary by device family.

Both the *Static* and the interface points between the *Static* and the *Reconfigurable Partition* need to be identical for all the *Reconfigurable Modules* in the design.

Vivado achieves this by preserving the *Static* implementation and reusing it to implement subsequent *Reconfigurable Modules*. An additional innovation in Vivado is the creation of virtual I/O for each of the interface port called a *Partition Pin*. *Partition Pins* can be locked to specific anchor points within the routing tiles and maintained across *Reconfigurable Modules*. This consumes no LUTs or flip-flops, thus reducing resource overhead and timing delays at the interface.

Vivado generates a *partial bitstream* file for each *Reconfigurable Module* in each *Reconfigurable Partition* as well as a *full bitstream* which contains the data for both the *Static* and the *Reconfigurable Module(s)* being implemented. The *full bitstream* is used for initial configuration of the FPGA, while the *partial bitstreams* are used for switching in and out the various *Reconfigurable Modules*. Loading of *partial bitstreams* into the FPGA is generally performed via the FPGA's standard external configuration ports or via the internal configuration ports which can be incorporated into the *Static* portion of the design.

Partial Reconfiguration takes advantage of the FPGA's addressable configuration infrastructure which allows specific areas of the FPGA to be reconfigured. The smallest addressable segment of the FPGA is known as a *Configuration Frame*. Each frame typically corresponds to a single column of resources which is a clock region in height. As such each frame contains a single resource type, for example, DSP, block RAM, CLB, or routing interconnect; the actual number of resources in each frame depends on the resource type and varies by device family.

19.1.3 Design Considerations

In order to take full advantage of the potential benefits of Partial Reconfiguration for a given application, you need to take on board a number of considerations prior to starting the design. These are divided into three different categories:

- FPGA device family
- Design structure
- Support functions for Partial Reconfiguration

19.1.3.1 FPGA Device Family

Vivado currently supports Partial Reconfiguration for all production devices for all families starting with 7-Series.

Starting with the UltraScale device family, all resources except the *configuration block* can be partially reconfigured, while support in 7-Series is limited to CLBs, DSPs, and block RAMs. As UltraScale devices allow IOs, BUFGs, MMCMs, and other clocking components to reside inside the *Reconfigurable Partition*, different clocking structures can now be supported inside any *Reconfigurable Module*. It should be noted that clocks sourced from within the *Reconfigurable Module* may only be used to clock logic inside that same *Reconfigurable Module. Reconfigurable Module* clocks cannot be used to clock logic in *Static*. The addition in UltraScale silicon of more granular control of global reset after Partial Reconfiguration has removed the 7-Series requirement for clock-region alignment for *Reconfigurable Partition* floorplans. It is, however, still recommended that the *Reconfigurable Partition* floorplan be a regular rectangle which aligns to device clock regions. With changes in the stacked silicon interconnect (*SSI*) in UltraScale, *Reconfigurable Partitions* and *Reconfigurable Modules* are now able to span multiple super logic regions (SLRs) and are no longer restricted to a single die as is the case with 7-Series *SSI* devices.

19.1.3.2 Design Structure

The most important design consideration is the choice of an appropriate instance on which a *Reconfigurable Partition* is set. This instance should be defined to incorporate the full functionality that is being reconfigured at a given time under a single hierarchical block. If the function being reconfigured is made up of several hierarchical blocks, these must all be merged under a single hierarchical block.

Ensure that the resources required by all the *Reconfigurable Modules* are reconfigurable for the device family being used. Therefore, the design should be structured in a way—such that resources that cannot be reconfigured reside in the *Static* portion of the design—outside the *Reconfigurable Partition*.

The ports of the instance selected will be the *Partition Pins* of the *Reconfigurable Partition*. These should be the union of the pins of all the *Reconfigurable Modules* associated with that *Reconfigurable Partition*.

Partial Reconfiguration is designed to support unconnected input and output *Partition Pins*. Unconnected output *Partition Pins* will be tied *high* by default. If you need to tie them to the *ground*, you need to do so explicitly in the *Reconfigurable Module*. However, it is worth noting that explicitly tying to the *ground* is resource inefficient. Creating a *Reconfigurable Module* where all inputs and outputs are unconnected results in a black-box module which can be used to *turn off* functionality inside the *Reconfigurable Partition*.

Optimization across the *Reconfigurable Partition* boundary is prohibited in order to avoid optimization of *Static* to suit one *Reconfigurable Module* at the expense of another. Therefore, ensure that the design does not rely on optimizations across *Reconfigurable Partition* boundary. In addition, logic upstream and downstream of unconnected *Partition Pins* does not get optimized away by Vivado.

As with any FPGA design, achieving timing closure is key, it is recommended that registers are inserted on both sides of each *Partition Pin*. Registers on both sides of the *Reconfigurable Partition* boundary allow the Vivado tools to maximize the timing budget when implementing the *Static* and each of the *Reconfigurable Modules*.

19.1.3.3 Support Functions for Partial Reconfiguration

In order to allow the Partial Reconfiguration process to operate correctly, a number of support functions need to be added to the design. These can reside in the *Static* design, the board or with the system in which the FPGA is being used. These include storing of partial bitstreams, triggering the Partial Reconfiguration process, delivering of the partial bitstreams to the FPGA's configuration memory, as well as decoupling the *Static* design from the *Reconfigurable Partition* during the Partial Reconfiguration process.

Xilinx provides a *Partial Reconfiguration Decoupler IP* which can be used to decouple the *Static* from the *Reconfigurable Partitions* and can be driven by the user's *Static* design. An alternative is to use an enable signal on the timing registers on the *Static* side of the design to decouple the design during Partial Reconfiguration.

The more *Reconfigurable Partitions* and *Reconfigurable Modules* a design contains, the more storage would be required to store the *partial bitstreams* generated by Vivado. *Partial bitstreams* can be stored in on-board nonvolatile memory or off-board on an external storage location. Regardless of where it is stored, the design requires a means of transferring these partial bitstreams from their storage location into FPGA's configuration memory.

The Xilinx *Partial Reconfiguration Controller IP* can be used to help manage the transfer of partial bitstreams into the FPGA's configuration memory. Section 19.1.5 gives more insight into *partial bitstream* handling, the FPGA's internal and external configuration ports, and the means by which the FPGA can be configured.

19.1.4 Design Tool Flow

The Vivado Partial Reconfiguration tool flow involves a number of simple steps:

1. Synthesize the *Static* with *Reconfigurable Partitions* as black boxes.
2. Synthesize each of the *Reconfigurable Modules* separately in *out-of-context mode. Out-of-context mode* synthesis results in a design being synthesized without IOB insertion, which allows it to be stitched into the rest of the design at a later stage. If IOBs are required inside a *Reconfigurable Module*, then these must be explicitly instantiated.
3. Create a physical area constraint or *pblock* to define the *Reconfigurable Region* for each *Reconfigurable Partition*. This area should contain all the resources required for each of the *Reconfigurable Modules* and will be used to contain all *Reconfigurable Module* routing. *Static* logic is excluded, while *Static* routing can enter this area.
4. Set *HD.RECONFIGURABLE* property on each *Reconfigurable Partition*.
5. Implement the *Static* with one *Reconfigurable Module* per *Reconfigurable Partition*. Save a copy of the fully routed design.

6. Remove *Reconfigurable Modules* from this design and save a static-only copy of the design. This copy will allow black-box partial bitstreams to be generated and used to remove logic from the *Reconfigurable Partitions* on the FPGA.
7. Lock the static placement and routing.
8. Add a different *Reconfigurable Module* to static-only design to each *Reconfigurable Partition*, implement, and save the fully routed design.
9. Repeat Step 8 until all *Reconfigurable Modules* are implemented.
10. Run Partial Reconfiguration verification utility on all routed designs.
11. Generate bitstreams for each routed design; this generates *Full Bitstreams* and *partial bitstreams* for each *Reconfigurable Module*.

Any of the *Full Bitstreams* generated can be used to initially configure the FPGA; the choice should be determined by the functionality required at the start of the system. The *partial bitstreams* for the *Reconfigurable Modules* that are generated are compatible across configurations; therefore, the *partial bitstreams* generated can be used with any full bitstream even if they were not generated as part of the same configuration.

19.1.5 Configuration Management

Storing and managing *partial bitstreams* is key to the success of Partial Reconfiguration in a design. Storage of *partial bitstreams* is typically outside the FPGA, either on a nonvolatile flash memory on the board or on another remote medium, and accessible to the FPGA via PCIe, Ethernet, or other data transfer protocol. Managing these *partial bitstreams* can be done using an external processor or an internal state machine or processor within the *Static* region of the FPGA. The processor or state machine determines which *Reconfigurable Module* should be loaded, where the partial bitstream for that *Reconfigurable Module* resides as well as when and how it will be downloaded into the FPGA's configuration memory. The Xilinx *Partial Reconfiguration Controller IP* can also be used to help manage partial bitstream configuration.

Depending on the location of the *partial bitstreams* and the management engine used, various configuration ports can be used to configure the FPGA. The following are the available configuration ports:

- *ICAP* (*internal configuration access port*): The primary choice where configuration management is being done internally to the FPGA. This requires a controller as well as logic to drive the ICAP interface.
- *MCAP* (*media configuration access port*): Provides access to configuration memory from one specific PCIe block only in UltraScale devices.
- *PCAP* (*processor configuration access port*): The primary configuration mechanism for Zynq-7000 SoC designs.
- *JTAG*: Test and debug port. Mainly driven by the Vivado Hardware Manager.

- Slave *SelectMAP* or slave serial: A good choice to perform full and partial reconfiguration, especially when using an external processor.

19.2 Tandem and Field Update

The PCI Express specification requires the PCIe link to be ready to link train with a peer within 120 ms after power is stable. This is nominally referred to as the *100 ms boot time*. Meeting this requirement is a challenge for large FPGAs due to the size of the bitstream and typical configuration rates available. *Tandem* support in 7-Series and UltraScale allows the PCIe to be up and ready to link train within the required timeframe.

19.2.1 Key Concepts

The Tandem flow allows the PCIe block in the FPGA to meet the 120 ms boot-up requirement by splitting the configuration into two stages:

- *Stage 1*: The minimum PCIe functionality needed to ensure device discovery is configured. This stage requires a very small bitstream that can be configured in much less than 120 ms and is capable of handling all transactions during enumeration time.
- *Stage 2*: The rest of the FPGA is configured with the user design after the PCIe block becomes active.

There are two tandem configuration methods supported, Tandem PCIe and Tandem PROM. Both methods employ the *two-stage* bitstream configuration principle outlined above. In both cases, *Stage 1* is configured via an on-board PROM which resides on the board, in order to meet the 120 ms start-up time. The main difference is in the delivery of the *Stage 2* bitstream; Tandem PROM uses the same on-board PROM, while in Tandem PCIe, the PCIe interface is used. Unlike Partial Reconfiguration, the Tandem approach never reconfigures a frame. Every frame in the device is configured only once. If dynamic updates to the user application are required, Partial Reconfiguration or the *Field Update* flow should be used.

The tandem with *Field Update* flow was introduced starting with the UltraScale architecture; Tandem configuration methods are used to initially configure the device when the power is turned on, followed by Partial Reconfiguration of the full *Stage 2* logic. Thus, the Field Update flow allows multiple *Stage 2* bitstreams to be downloaded on demand, without the need to reconfigure the *Stage 1*, thus maintaining the PCIe linkup throughout. Figure 19.5 shows how the Tandem PROM, Tandem PCIe, and Tandem with *Field Update* flows operate.

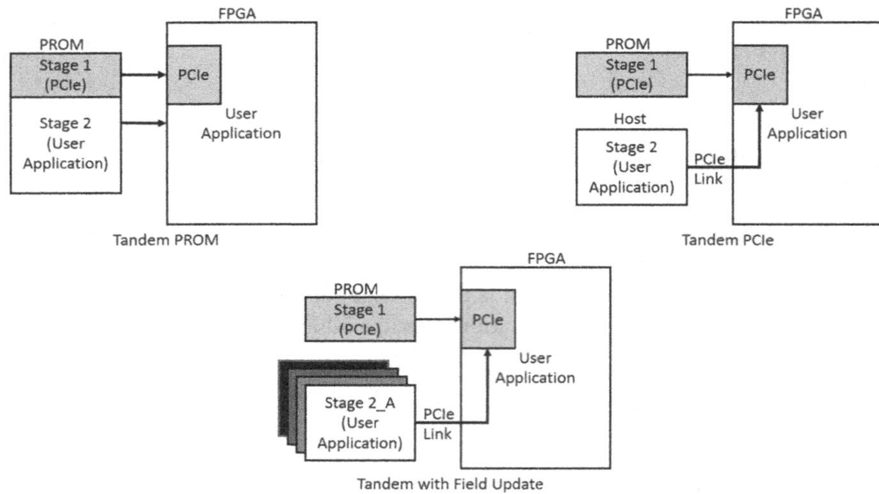

Fig. 19.5 Tandem PROM, Tandem PCIe, and tandem with Field Update configuration flows

19.2.2 Design Tool Flow

The support for the tandem and tandem with Field Update flows is embedded within the PCIe core. The PCIe core and example design should be used as the foundation of any applications that utilize these flows. The following steps outline the tool flow to be followed by you:

1. Select the type of tandem flow required and generate the core.
2. Open the example project, and implement the example design.
3. Use the IP and XDC from the example project as the basis of your project.
4. Synthesize and implement your design.
5. If using tandem with Field Update, follow steps 6–10 from Sect. 19.1.4.
6. Generate bitstream and PROM files required.

19.2.3 Configuration Management

Tandem PROM and Tandem PCIe flows both rely on initial PROM configuration of *Stage 1* followed by *Stage 2* being configured via the external configuration pins in Tandem PROM or via the PCIe link in Tandem PCIe.

In Tandem PCIe, the PCIe IP provides an internal interface to the configuration memory. In 7-Series this is achieved by an explicit connection to the *ICAP* (internal configuration access port). This connection is disabled after *Stage 2* configuration. In UltraScale the connection to the configuration memory is made via the *MCAP* (media configuration access port) which is embedded inside the PCIe block. This connection remains enabled even after *Stage 2* configuration is complete. Access to the *MCAP* after *Stage 2* is the key enabler for the Tandem *Field Update* flow.

19.3 Hierarchical Design (HD) Preservation

Hierarchical design (*HD*) flows enable you to partition a design into smaller modules that can be implemented independently, before choosing whether or not to reuse the results at the top level of the design.

19.3.1 Key Concepts

Hierarchical design flow provides the ability to take a given module, synthesize and implement it independently, and then reuse the results in an overall design. There are two parts of the hierarchical design flow: *Module Analysis* and *Module Reuse*.

In *Module Analysis* you can synthesize, implement, and conduct resource or timing analysis on a module without the need of special wrappers. The implementation is done with no *IOs* or *clocks*. These need to be explicitly specified if needed. The implementation results can then be saved for reuse.

In *Module Reuse* you take the results of an implemented *Module Analysis* run, lock-down, and reuse them in a top-level design. There are two variants of *Module Reuse*: *bottom up* and *top down*.

Bottom-up reuse is where you ran the *Module Analysis* flow without prior knowledge of the top-level design. This allows you to reuse the same *Module Analysis* results for multiple top-level designs on the same device.

Top-down reuse is where you use a top-level design and floorplan to generate *out-of-context* constraints, to be used by independent *Module Analysis* runs, before reusing the results to assemble the top-level design. This flow allows a team to work simultaneously on portions of the same design.

19.3.2 Design Tool Flow

The Vivado tool flow for hierarchical design is split into *Module Analysis* and *Module Reuse*. To run the *Module Analysis*, use the following steps:

1. Synthesize the module or IP in *out-of-context* or *bottom-up synthesis*.
2. Set the *HD.PARTITION* property on the module.
3. Add clock and timing constraints specific to that module.
4. Floorplan the area into which the module will be placed.

5. Add *out-of-context* constraints including *HD.CLK_SRC* property as well as *partition pin* locks and optimization constraints.
6. Implement the module and save the placed and routed module results.

To run the *Module Reuse* flow, use the following steps:

1. Synthesize the top level with black boxes for module instances.
2. Set *HD.PARTITION* property on the module instances.
3. Read in results from *Module Analysis* run, into the relevant instances.
4. Lock the implementation results of the modules that have just been read in. This can be done at either logical, placement, or routing level.
5. Implement the remainder of the design.

19.4 Isolation Design Flow

The *Isolation Design* flow was developed to allow independent functions to operate on a single chip with the sufficient level of isolation required for various certifications. Applications of this flow include redundant type I cryptographic modules or resident safety-critical functions.

19.4.1 Key Concepts

There are a few unique design details that you must adhere to, in order to achieve an FPGA-based *isolation design* flow solution. The requirements that a design needs to meet in order to take advantage of the *isolation design* flow are shown in Fig. 19.6 and include:

- *Isolated Module*: Each function to be isolated must be in its own level of hierarchy and reside within its own physical region of the FPGA.
- *Fence*: This is a set of unused tiles with no logic or routing used—to separate the *isolated modules*. This has to be a minimum of one non-routing tile in depth.
- *Trusted Routing*: On-chip communication between isolated functions is achieved through the use of *trusted routing*. Vivado chooses one to one routes along the coincident physical borders of *isolated modules*.
- *Top Level*: Only global logic including BUFG and MMCM is allowed at the top level. All other logic must reside inside an *isolated module*.
- *IOBs*: IOBs can be instantiated or inserted inside the *isolated modules*.

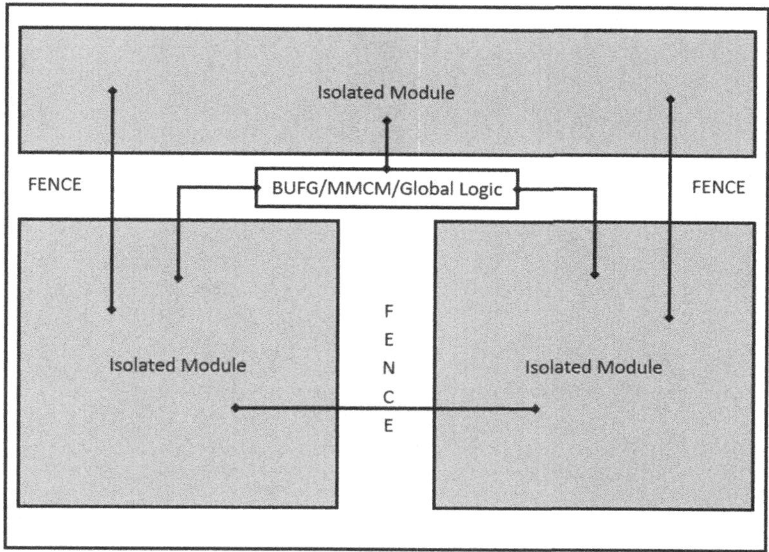

Fig. 19.6 Isolated design flow floorplan with trusted routes and fences shown

19.4.2 Design Tool Flow

The *isolation design* flow relies on you logically partitioning the design such that each *isolated module* resides in a different hierarchical block directly under the top level of the design. Once this is achieved, there are a few steps that you need to follow:

1. Set the *HD.ISOLATED* property on each *isolated module*.
2. Set the *HD.ISOLATED_EXEMPT* property on any logic at the top level.
3. Synthesize the design.
4. Floorplan the *isolated modules*.
5. Run isolation verification on the floorplan to ensure adequate fencing.
6. Implement the design.
7. Run isolation verification on routed design to ensure correct isolation.
8. Generate bitstream.

References

(A) **Xilinx User Guides, Tutorials, Product Guides, Application Notes, White Papers etc.**

Xilinx keeps updating its documents based on the last released version of the Vivado software tool. You should refer to the document corresponding to the version number of the Vivado software being used for your design. Similarly, you should refer to the documents for the specific Silicon architecture that you are using.

(1) UltraScale Architecture-Based FPGAs Memory Interface Solutions LogiCORE IP Product Guide (PG150)
(2) LogiCORE IP UltraScale FPGAs Gen3 Integrated Block for PCI Express (PG156)
(3) LogiCORE IP System Management Wizard Product Guide (PG185)
(4) Equalization for High-Speed Serial Interfaces in Xilinx 7 Series for 7 Series FPGA Tranceivers (WP419)
(5) Leveraging 7 Series FPGA Transceivers for High Speed Serial I/O Connectivity (WP431)
(6) Xilinx Power Estimator User Guide (UG440)
(7) Leveraging UltraScale Architecture Transceivers for High Speed Serial I/O Connectivity (WP458)
(8) 7 Series FPGAs GTH/GTX Transceivers User Guide (UG476)
(9) 7 Series FPGAs GTP Transceivers User Guide (UG482)
(10) UltraScale Architecture Clocking Resources User Guide (UG572)
(11) UltraScale Architecture Memory Resources (UG573)
(12) UltraScale Architecture Configurable Logic Block User Guide (UG574)
(13) UltraScale Architecture GTH Transceivers User Guide (UG576)
(14) UltraScale Architecture DSP Slice User Guide (UG579)
(15) UltraScale Architecture System Monitor (UG580)
(16) Zynq-7000 All Programmable SoC Technical Reference Manual (UG585)
(17) Driving the Xilinx Analog-to-Digital Converter (XAPP795)

S. Churiwala (ed.), *Designing with Xilinx® FPGAs*,
DOI 10.1007/978-3-319-42438-5

(18) Vivado Design Suite User Guide: Vivado TCL Commands (UG835)
(19) Vivado Design Suite Tutorial: High-Level Synthesis (HLS) (UG871)
(20) UltraScale Architecture and Product Overview (DS890)
(21) Vivado Design Suite User Guide: Design Flows Overview (UG892)
(22) Vivado Design Suite User Guide: Using the Vivado IDE (UG893)
(23) Vivado Design Suite User Guide: Using Tcl Scripting (UG894)
(24) Vivado Design Suite User Guide: System-Level Design Entry (UG895)
(25) Vivado Design Suite User Guide: Designing with IP (UG896)
(26) Vivado Design Suite User Guide: Model-based DSP Design using System Generator (UG897)
(27) Vivado Design Suite User Guide: Embedded Hardware Design (UG898)
(28) Vivado Design Suite User Guide: Logic Simulation (UG900)
(29) Vivado Design Suite User Guide: Synthesis (UG901)
(30) Vivado Design Suite User Guide: High-Level Synthesis (HLS) (UG902)
(31) Vivado Design Suite User Guide: Using Constraints (HLS) (UG903)
(32) Vivado Design Suite User Guide: Hierarchical Design (UG904)
(33) Vivado Design Suite User Guide: Design Analysis and Closure Techniques (UG906)
(34) Vivado Design Suite User Guide: Power Analysis and Optimization (UG907)
(35) Vivado Design Suite User Guide: Programming and Debugging (UG908)
(36) Vivado Design Suite User Guide: Partial Reconfiguration (UG909)
(37) Vivado Design Suite User Guide: Getting Started (UG910)
(38) Vivado Design Suite Tutorial: Programming and Debugging (UG936)
(39) UltraFast Design Methodology Guide for the Vivado Design Suite (UG949)
(40) Vivado Design Suite Quick Reference Guide (UG975)
(41) MicroBlaze Processor Reference Guide (UG984)
(42) Vivado Design Suite User Guide: Designing IP Subsystems Using IP Integrator (UG994)
(43) Zynq UltraScale+ MPSoC Technical Reference Manual (UG1085)
(44) Using Tandem Configuration for PCIe in the Kintex-7 Connectivity TRD (XAPP1179)
(45) UltraFast High Level Productivity Design Methodology Guide (UG1197)
(46) Isolation Design Flow for Xilinx 7 Series FPGAs or Zynq-7000 AP SoCs (Vivado Tools) (XAPP1222)

(B) **Other References**

(1) Virtual Wires: Overcoming Pin Limitations in FPGA based Logic emulators. http://www.princeton.edu/~mrm/ee470/fccm93.pdf
(2) High-Speed Serial I/O Made Simple: A Designers' Guide, with FPGA Applications. http://www.xilinx.com/publications/archives/books/serialio.pdf
(3) Synopsys Certify tool Overview. http://www.synopsys.com/Prototyping/FPGABasedPrototyping/Pages/Certify.aspx

(4) FPGA-based Prototyping Methodology Manual: Best Practices in Design-for-Prototyping. Synopsys and Xilinx
(5) Three Ages of FPGAs: A Retrospective on the First Thirty Years of FPGA Technology: By Stephen M. (Steve) Trimberger, Fellow IEEE; Vol. 103, No. 3, March 2015 | Proceedings of the IEEE
(6) DSP: Designing for Optimal Results High-Performance DSP Using Virtex-4 FPGAs; http://www.xilinx.com/publications/archives/books/dsp.pdf
(7) Field-Programmable Gate Array Technology; Stephen M. Trimberger – Editor; Springer Science & Business Media, Jan 31, 1994; http://www.springer.com/in/book/9780792394198
(8) DDR4 documentation from JEDED. Registration required. https://www.jedec.org/standards-documents/results/jesd79-4%20ddr4
(9) RLDRAM-3 spec from Micron. https://www.micron.com/products/dram/rldram-memory
(10) QDRIV spec from Cypress. http://www.cypress.com/products/qdr-iv
(11) Simulink and MATLAB product descriptions. www.mathworks.com
(12) Constraining Designs for Synthesis and Timing Analysis; Gangadharan Sridhar, Churiwala Sanjay; Springer Science and Business Media, 2013; http://www.springer.com/us/book/9781461432685

Index

S. Churiwala (ed.), *Designing with Xilinx® FPGAs*,
DOI 10.1007/978-3-319-42438-5

Printed in the United States
By Bookmasters